"Endless invention, endless experiment,
brings a knowledge of motion,
but not a stillness of wisdom."

T. S. Eliot

Disclaimer & Copyright:

This is not an official company publication and no product endorsements are offered or implied. We recognize further that some model names and designations mentioned herein are the property of the trademark holder.

The information in this book is true and complete to the best of our knowledge. All recommendations are made without any guarantee on the part of the authors, who also disclaim any liability incurred in connection with the use of this data or specific details.

ISBN-13: 978-0-9843442-3-9

Front and back cover graphic design and layout by:
Michelle Giacchetta
President of MAG Designs Syracuse, New York

Published by:
Buffalo Road Imports, LLC
10120 Main Street
Clarence, NY 14031, USA
www.BuffaloRoadImports.com

Printed in USA

Northwest Engineering Company

A Photographic Archive collection

Volume Two

The Classic years 1941 to 1966

Photographic compiling, text, and editing by: M.E Folsom & M. J Torres

The decade of the 1950s would prove to be the height of production for Northwest Engineering Company. The company would produce more machines during these ten years than in any other decade of its existence. A total of over 8,000 units would come out of the factory located in Green Bay, Wisconsin during those ten years. For the years referenced in this work, 1940 to 1966, Northwest would produce a total of over 18,500 machines. The scene above shows a bevy of newly assembled Model 25-D cranes, shovels, and pullshovels, with a few other model types in between at the height of Northwest's production. The Model 25 would be the most prolific unit ever produced by Northwest, with over 5,200 being delivered to customers worldwide.

(Front cover) A "Classic years" Northwest Model 95 pullshovel is seen loading out, by today's standards, a rather small capacity ten-wheel dump truck. The machine with its 3 cubic yard capacity bucket was the company's largest unit for many years.

Table of Contents

Introductions **7**

Acknowledgements **9**

The Northwest Engineering Company "Classic Years" **11**

The Third and Fourth Generation Machines **14**

- Models 15, 18 and 18-D **15**
- Model 20 .... **29**
- Model 25 .... **32**
- Model 25-D **51**
- Model 26 .... **58**
- Model 40 .... **60**
- Model 41 .... **62**
- Model 6 Updates **78**
- Model 70 .... **94**
- Model 71 .... **96**
- Model 78-D **99**
- Model 80 and 80-D **104**
- Model 85 .... **142**
- Model 95 .... **146**
- Model 180-D **164**
- Model 190-D **177**
- Views of things to come **193**

Appendix: **197**

Murphy Diesel Engines
Northwest C.I.M.A. (road show) images

Introduction

Webster offers up a plethora of definitions that pertain to the word "**classic**". One is "something noteworthy of its kind and worth remembering", another is "a work honored in its field" and yet another is "a traditional definitive standard of the highest quality'. All of these delimitations can apply to the machines produced by Northwest Engineering Company of Green Bay, Wisconsin between the years 1941 to 1966.

This photographic archive is the second in a trilogy on Northwest and the magnificent machines that the company produced. This volume pertains to the "**classic years**" of engineering, designs, and manufacturing of full revolving fiction-based construction and mining equipment of Northwest. The firm would reach its zenith in both popularity of its machines and sales; this had as much to do with the company's rock solid reputation for building a superior product as well as the correct set of circumstances in the United States and world economy.

The post-World War Two war years in the United States and abroad created a construction boom unparalleled ever before seen in the history of mankind. America and its new found position in the world brought forth by its returning servicemen ushered in what is now commonly known as the baby boomer era, bringing with it a tremendous need to build and to fill growing demands for housing and a want for a better life. It should be noted that many veterans returning from this titanic conflict would find themselves in the seat of a Northwest machine in their post-war careers. The training that "Uncle Sam" gave, whether it was from the Navy Seabees or an Army Engineering unit, certainly qualified these men more so to be an operating engineer of the utmost skill.

The Northwest Engineering Company's classic years of production brought forth a refinement of their engineering and designs that devotees and enthusiasts of the company and its equipment have come to admire more so than any other period.

Streamlined houses, large welded attachments, rugged digging implements and of course, pekin orange and black color schemes, all of these identifiers would define the "classic years" for Northwest Engineering Company and its machines.

As stated in the previous volume presented by Mario Torres and myself, a few of the images shown here within have been used before in our very first publication regarding Northwest Engineering. The reasons are two fold, first to omit them would narrow the oeuvre and they are needed to tell the story. They are rare and in some cases the only photographic documentation in existence to express that particular part of the narrative. The second reason for the reprinting of certain images is to give an improved presentation for them. Mr. Torres and I realized when we presented them in our very first work that the quality in a few of the images was at best marginal. These images were scanned from original prints that, to our knowledge, are the only copies in existence. The images represented here and in the previous volume have been cleaned-up and enhanced, but have not in any way been altered to change content. The photographs we now present within this work should also quell any detractors and or reviewers of that previous first work.

The structure of this work will contrast slightly from the first volume presented in that there will be no defining chapters between what we the authors call the third and fourth generation machines of Northwest. (These machines never received a true heading for these generations.) The reason for this is because the evolution process between the two contemporaries is not as well defined as the first and second generational forays. This is not to say that these two latter generations were blurred, but many models were introduced and upgraded at different times. No definitive date separates the two.

1940 to 1966 saw a tremendous evolution at Northwest Engineering Company; these years would present some of the finest and most respected machines in the material handling field, making the company one that other manufacturers would aspire to.

Thus the saga continues,

Enjoy,

This opus is dedicated to all the veterans of the armed services of:

The United States of America

Let it be known that their service and ultimate sacrifices will never ever be forgotten.

Acknowledgments

Works of a photographic nature obviously require images of very good quality. Some of the photographs used in this second volume, as with volume one, came from our good friend Mr. Alan Brickett who as always had a large assortment to ponder over. Again Alan, we so graciously thank you for sharing these rare finds from your archive with us. The other large collection to wade through and use came from yet another Northwest engineer Mr. Bradly Knistern. Mr. Knistern was not only a mechanical engineer with the company, but also a bit of an amateur photographer who documented over the years (1936 to 1969) many of the prototypes, unique machines and attachments at the factory. This collection of approximately 650 small formatted black & white photographs bound in two hard cover volumes was purchased sometime in the early 1990s by our good friend Mr. Horace Lucas of Green Bay to add to his growing archive regarding Northwest. This collection was in turn purchased by Mr. Everett DeBerry of Deer Park, Maryland, sometime around 2006. In 2012 Mr. DeBerry offered the entire Lucas collection to its present owner, author M. E. Folsom. Certainly it has been a long journey to get all the above material properly published.

The next major contributor of imagery to this endeavor is a new acquaintance, Mr. David Pickhardt of Randolph, Wisconsin. While doing the research for the first book released, we came in contact with Mr. Pickhardt, who so generously lent us his entire *Material Handing Illustrated* collection for use. This collection filled in a few voids of our own that for years always seemed to elude us in our collecting of Northwest memorabilia. Thank you Dave, your collection is really a treasure trove. Another contributor, Mr. James Hilgartner of Maryland, provided a vast array of Murphy Diesel engine information and images from his archive.

An important person to lend a hand in getting this effort to print is our good friend and fellow author Mr. Edgar Browning, who we imagined suffered long hours with the scanning process. Mr. Browning somehow found the time to do our job as well as keep his own work to a stringent schedule. If you see any of Mr. Browning's works on display documenting the photographic state histories of road building; please procure yourself a copy or copies, you won't be disappointed. Again, thank you Edgar.

A special thank you goes to renowned author Mr. Eric Orlemann who pointed us in the right direction with the proper publication of this series.

Mr. Brandon Lewis, our publisher proper, thank you for taking a chance with us on the previous photographic volume, this present one, and the next. You keenly saw what others in your field couldn't. And finally Shawn Lewis, Brandon's brother, thank you for all your wonderful work on the layouts and photograph restorations associated with putting together a work of this nature. Your work truly made the difference.

With friends and associates like this helping out, how can we go wrong!

M.E. Folsom & M. Torres

Views of Northwest Engineering Company factory during its "classic years" in Green Bay, Wisconsin. The above photograph of the main plant to the south of W. Walnut Street was taken in 1940. The bottom photograph of the cab assembly building and the north storage yards was taken in 1957.

The Northwest Engineering Company

"Classic years"

The "classic years" of Northwest Engineering Company's machines had its genesis in the decade of the 1930s. As the second generation machines were taking hold of their market, chief engineer Paul Burke was also hard at work with developing new ideas and designs. As his work load increased, it was realized that an assistant engineer would be required to handle the additional tasks. Edward Blanchard had come on board with Northwest in the late 1920s as a mechanical engineer, graduating from the University of Michigan with a Bachelors of Arts in this field. Burke initially noticed a keen and capable mind in Blanchard and charged him with the task of machine failure testing and analysis. Blanchard went about setting up a series of jigs and fixtures to simulate the forces imposed on a machine from the work it was expected to do. These quite complex set-ups tested attachment design and component stoutness through a system of mechanical manipulation by winches with sheaves, cables and turnbuckles. The testing and analysis done on the attachment undoubtedly made the Northwest product the superior machine we all have come to know.

The three photographs above show the type of testing Edward Blanchard developed at Northwest for attachment strength and potential failure. This is a 1 cubic yard front shovel attachment for a Model 40 developed in the mid 1930s. Blanchard would gradually introduce the single stick shovel attachment across Northwest's line.

The above photographs show fatigue testing of the crawler side frames. The white paint applied to sections of the track frames are meant to easily show stress marks as pressure is applied.

The other major development Edward Blanchard brought to Northwest was electric arc welding. This new practice of structural steel fabrication was initially developed in the mid-1920s and by the mid-1930s had been perfected. By bringing in welding as the main fabrication method, production costs would be substanually reduced.

Blanchard would also be responsible for the change in appearance of the main digging attachment offered by Northwest, the front shovel, starting with the newly designed lower end capacity machines of the now third generation. The front shovel would become a split boom with a single stick of welded construction and much simpler cable dual crowd, replacing the previous twin-stick design. Pullshovels received the same treatment from a riveted construction to welded fabrications.

This design criteria for attachments would slowly be introduced on up the line in the following decades to come.

By 1950, Vice President of Engineering Paul Burke would decide to retire. The position of chief engineer would naturally fall to Edward Blanchard. Burke hand picked his successor to take over the helm of the department. From 1950 to his retirement in 1964, Edward Blanchard was the man who directed Northwest though its "Classic years". He worked to update the models as the industry changed, bringing in a modern look to the machines with periodic cab upgrades, but most importantly, he strived to constantly improve the attachments for these machines. The most notable of these cab revisions resulted in the famous "breadbox" house. The "classic" cab would feature lines that would be common to the era with a rounded back, rounded roof corners, and sloped blindsides for better operator cross vision. It really did have a mid-1950s standard American breadbox look. The famous color scheme of pekin orange and gloss black that had become the main identifier for all Nortwests saw its inception during the early combined Burke and Blanchard years of the 1930s and 1940s. Although both would not be the responsible party for this marketing stratagem, both did see a need to revitalize Northwest's appearance.

Aesthetics and attachments would not be the only things upgraded thoughout these years. The internal workings of all Northwest equipment saw their share of improvments too. From the gantries down to the draw-works, through the clutches and brakes, and finally in 1962 with the introduction of what would make operating a Northwest machine even easier than Paul Burke imagined, air controls. Known as "Cushion-Air", it would set the standard for all of Northwest's traditional machines until the end of production in 1990.

1962 would also mark year that Northwest would produce their largest ever machine line with introduction of the Model's 180-D and 190-D . This substantial jump in machine size was warranted by the constant request of customers wanting a larger capacity machine for their needs.

All of these improvements that Edward Blanchard would oversee as Northwest's chief engineer during his reign would define what would become known as Northwest's classic years. With the death of Northwest's president L.E. Houston on September 6, 1966, this chapter in Northwest's history would close for the company. The winds of change were starting to blow through the company in their leadership, engineering, and sales organizations and also the industry for which Northwest catered to.

The Northwest sales literature above is a fine example of the transitional stage in the company. This brochure, dated 1944, shows that the company was now producing eighteen different size pieces of heavy equipment, from the 3/8 cubic yard Model 15 to the unmentioned 3 cubic yard model 95. What is interesting is that on the right hand side, images of the company's second generation machines are presented with a Model 8 crane on the top and a Model 7 pullshovel on the bottom. The left hand side photographs show the newer Model 80-D shovel on top and a Model 95 dragline on the bottom. This is a good illustration of how the styles between machine generations were somewhat muddled during this time.

The Models 15, 18, & 18-D

Produced from 1936 to 1943 both models shared the same base machinery only differing in cubic yardage and capacity with the installation of different size dipper and booms.

The Model 15 had a 3/8 cubic yard capacity and a lifting capacity @ 10ft. of 8500 lbs. with a weight of 22,000 lbs. Only offered with a gasoline power-plant.

The Model 18-D had a ½ cubic yard capacity and a lifting capacity @ 10 ft. of 12,000 lbs. with a working weight of 24,000 lbs. Only offered with a gasoline power-plant.

The smallest machine that Northwest Engineering Company would offer in their line, the 3/8 cubic yard capacity Model 15. Yet as small as it was, it offered many of the proven features of its larger brethren. Even at its small size, it still offered the Feather-Touch clutch control, independent cable crowd, and single stick/split boom shovel front.

A nice detailed view of a Model 15 shovel is seen here at the Northwest factory in Green Bay. The Model 15 had all the newest innovations brought forth by Paul Burke and Edward Blanchard. The operator seen here seems quite happy to demonstrate its features.

The cover for the first sales literature catalog of the new Northwest Model 15, dated 1937.

The next new ½ cubic yard shovel was the Model 18-D. The model shared all the same componentry of the Model 15, only differing in bucket and boom sizes for increased capacities along with an option for wider track tumbler pads, such as this unit is carrying. This particular unit is working in a salt warehouse. Check out the motorized buggies for hauling the salt from the loading area.

Northwest offered contractors a convenient means for transporting their Model 18 shovel, simply called The Northwest Trailer. A simple I-beam construction, with dual wheels and a detachable towing tongue to allow loading, it provided for speedy relocation from job to job.

A Northwest Model 18 climbs up on the blocks to be loaded onto the Northwest Trailer, demonstrating the relative ease of loading and unloading a machine for travel.

A more practical way to get a small shovel from jobsite to jobsite. This Model 15 shovel mounted on a chain drive, single axle Mack AC model chassis is motoring down a residential neighborhood; no doubt the youngsters of the area were interested in it when it arrived at its jobsite.

A shovel parked in front of the same house on a residential street as in the previous pictures, a dream come true for many a little boy back in the 1930s. Notice the additional reinforcing necessary for mounting the shovel. Could it be, that the operator/owner was so proud of his machine, that he drove it home every day, to park it proudly in front of his house?

A truck mounted Model 15 shovel is seen excavating a home building site. Truck mounting was a good way to enhance mobility between assignments, and increase machine utilization. A cab on this carrier, however, would have been better for the driver's protection against the elements!!

A truck mounted Model 18-D shovel. To mount a Model 18-D on a truck required a chassis of larger weight capacity. This dual tandem drive truck certainly had the proper factory requirements to support the slightly heavier Model 18-D shovel, and the carrier has a cab.

NORTHWEST
NORTHWEST

NORTHWEST

Previous page plus above picture: These three pictures depict truck mounted Northwest cranes on Mack AC single axle carriers. Prior to integrated, purpose-built carriers that would come later, off-the-shelf commercial trucks were modified for use as carriers. Frames were strengthened, and outrigger boxes were added. The Mack AC line was a popular candidate for conversion, as evidenced by the bottom picture on the previous page. Hard rubber wheels were the order of the day.

A Model 15 crane with clamshell bucket mounted on a half-cab carrier. Clamshell buckets were a popular accessory for a crane for excavating. Notice the triple hard-rubber wheels on the carrier. Although the outrigger beams are extended, the screw jacks or blocking placed beneath the beams that were common during this era do not appear to be in place. It wouldn't be a good idea to boom out or swing out without the jacks or blocking installed!

A Model 18 crawler crane working off a rail flatcar installing new rail over freshly laid ties. This was certainly a vast improvement over *"gandy dancing"* rails the traditional way.

The Model 20

Produced from 1937 to 1943
½ cubic yard capacity, 15,000 lbs. lift rating @ 10 ft. radius

A Model 20 shovel excavating for a basement foundation in the machine's hometown of Green Bay, Wisconsin. The nice, high reach enabled loading the truck at sidewalk level.

Two sequential view of the same Model 20 shovel from the previous page. The Model 20 was one of the more rare machine models produced by the company. The exact factory production numbers are unavailable. This one has done a fine job of excavating this basement, while spectators take in the action.

A Model 20 crane equipped with a clam works for Macco Construction Company at a crude oil field drilling operation located in Clearwater, California.

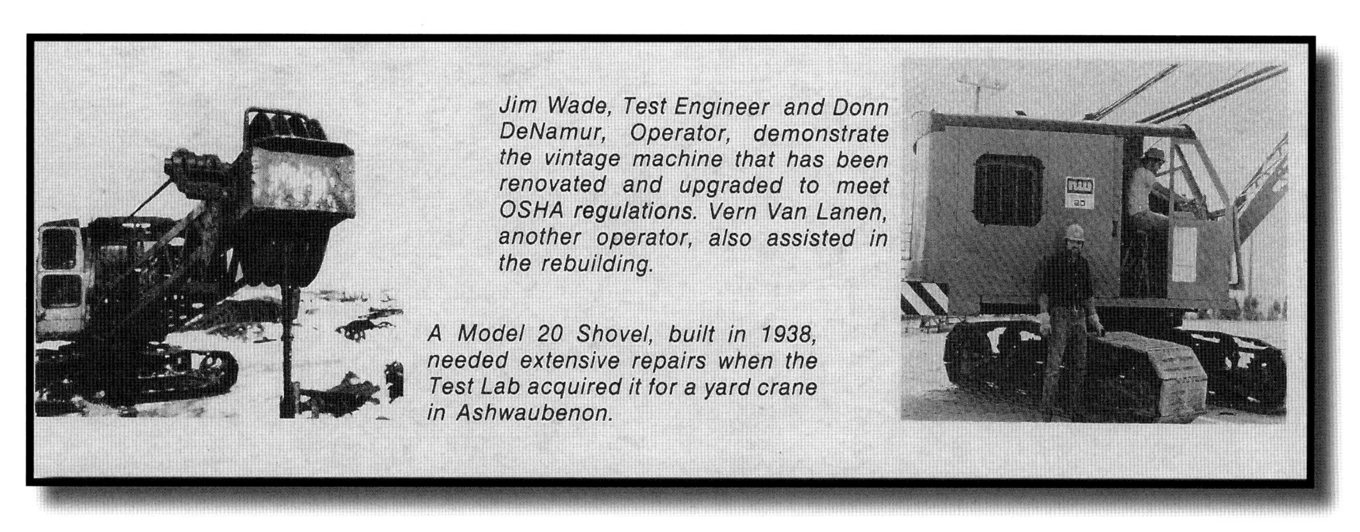

Jim Wade, Test Engineer and Donn DeNamur, Operator, demonstrate the vintage machine that has been renovated and upgraded to meet OSHA regulations. Vern Van Lanen, another operator, also assisted in the rebuilding.

A Model 20 Shovel, built in 1938, needed extensive repairs when the Test Lab acquired it for a yard crane in Ashwaubenon.

These two images were taken from fall 1978 Northwest employee's club magazine "Scope". This most likely was the last ever Model 20 working for the company.

The Model 25

Model 25 Produced from 1937 to 1958

Model 25-D produced from 1958 to 1974

The most popular small machine produced by Northwest was the Model 25. It sold in great numbers, and was equipped with all of the front-end attachments available: crane, dragline, shovel, and pullshovel. The very first production Model 25 was set up for dragline work. The early Model 25 machines had a more squared off machinery house.

Two views of the prototype Model 25's undercarriage. The basic design of the undercarriage was carried through all of the model lines up to the largest that Northwest would offer. In evidence here are the I-beam axles extending to the I-beam crawler frames, with a cast carbody bolted to the axles.

One of the first of batch of Model 25 shovels seen here at the factory in December of 1937.

One of the first Model 25 pullshovels in service, seen working for Central Contracting Company of Oshkosh, Wisconsin.

Top picture, L.E. Hoadley's shovel appears to dwarf the truck being loaded, and below, John C. Cashman's pullshovel opens a trench on a sewer job.

Similar to the Northwest Trailer offered for the Model 15 and 18, Northwest offered a larger version for transporting the Model 25. It utilized similar I-beam construction with a detachable towing tongue. Due to the heavier Model 25, additional rear axles and tires were fitted.

Following page:
Top picture: More *"gandy dancing"* with a Northwest, this time a Model 25 is being utilized for the arduous task.
Bottom picture: Another Model 25 equipped with a boom and clam places new ballast.

B&ORR
CC12
NORTHWEST
B & O 109097

The Edward G. Budd Company was a manufacturer of passenger railroad rolling stock in Philadelphia, PA. They utilized a Northwest Model 25 crawler crane with a gooseneck boom for lifting duties around the plant. Here, a truck frame needed moving. A coach for the Rock Island sits on the track alongside.

Following page: Two Model 25 cranes mounted on different style wagon carriers. Such wagon-mounted units were especially handy in the railyards.

NORTHWEST

NORTHWEST
ENGINEERING CO
CHICAGO ILLINOIS.
REPAIR PARTS

Another Model 25 crane doing its intended task of setting pipe. A self-propelled wagon carrier was an ideal machine for the urban pipeline contractor. Quickly movable on the job from place to place along the pipeline, it was ideal for fast laying of pipe.

This half-cab tandem-axle Mack is the carrier for the Model 25 crane owned by F. A. Canuso & Sons of Philadelphia, PA. It could be a bridge under construction and concrete is being poured; a bucket is being filled behind the paver as the crane waits to hoist and swing it to the pouring location.

NORTHWEST
THOMAS
PROCTER
CO. INC.
LONG BRANCH
NEW JERSEY
PHONE 844
C N J
84 700
MODEL 25
Inc.

NORTHWEST
THOMAS PROCTER CO. INC.
LONG BRANCH, NEW JERSEY
PHONE 844
C N J
84 087

Oilfield Trucking Company of Bakersfield, CA, used this Model 25 for clamming duties in the oilfields. It is mounted on yet another tandem-axle Mack carrier.

Previous page: A rather tight location, and the rear of the crane just clears the building behind it, as this Model 25 loads stone into railcars. Note the blocks beneath the extended outrigger beams on the carrier; outriggers were manually extended and retracted and either blocking or a screw jack was used to provide stability. This was adequate for the short booms used during the period.

Two sequence photographs of a Model 25 truck crane working with a log boom loading out for Schmidt Brothers of Greer, Idaho. Note the more modern carrier.

Another view of Schmidt Brother's Model 25. The customer-made, 30 foot long log boom is quite a stout piece of homemade engineering.

Integrated carriers were beginning to make their appearance. These purpose-designed carriers were built from the ground up to be dedicated carriers with stiffer, stronger frames to resist twisting, as opposed to off-the-shelf trucks that were previously modified for such service. This Model 25 truck crane owned by Ashdale Corporation charges hoppers with a clamshell bucket. Manually operated outriggers were still common.

Following page: Two more pictures of the Model 25 truck crane with integrated carrier. This unit displays a clean, purposeful appearance.

NORTHWEST

NORTHWEST

NORTHWEST
NORTHWEST

NORTHWEST

Claude C. Wood Company's Model 25 truck shovel is busy loading trucks with loose material, presumably on a road job.

Previous page: Along with the new carrier, the Model 25 cab sheet metal underwent an upgrade. It took on a more rounded appearance, resembling a breadbox. Although the front of the cab has been redesigned, the rear has yet to be revised. This particular unit goes quite nicely with its carrier, giving the machine a polished appearance.

Pullshovels weren't neglected and also got the truck-mounted treatment, and here, a Model 25 pullshovel mounted on a Dart 6X4 carrier works in what looks like rocky material. The operator got his exercise during the day jumping off the machine and into the carrier cab to move forward, and back into the machine as the ditch progressed.

An advertisement for the Model 25-D shovel in the 4th quarter issue of *Material Handling Illustrated*, dated 1966. As wheel loaders began making inroads into the construction scene, Northwest promoted the shovel's advantages over a wheel loader in condensed form in this ad, and detailed the advantages further in the booklet "You can do it better with a shovel".

Another advertisement from the same *Material Handling Illustrated*, but this time depicting a Model 25-D shovel and Northwest's history with the John Klemmer Company of Owatonna, Minnesota. The company's first Northwest machine, seen in the inset, was a Model 105 shovel, purchased in 1926, which was Northwest's second machine introduced as a shovel in 1922.

These handsome shovels have the fully-redesigned "breadbox" cabs, with the distinctive rounded rear, and square opening with rounded corners, topped off with the rounded roof. The "Northwest" plaque below the opening completes the classic look. Seen at the Green Bay factory, the 25-D in the top picture and the same machine seen below has as a special option, foldable catwalks. The rear of the machine seen to the left of the 25-D in the bottom picture is a Model 6 shovel.

Company advertisement for the Model 25-D pullshovel as it appeared in the 4th quarter 1961 issue of Northwest's Material Handling Illustrated. The rugged little 25-D pullshovel was a fine performer for the pipeline industry.

Model 25-D pullshovels performing trench excavations on a road job above, and a sewer job below. Many contractors throughout the 1950s and 1960s used a Model 25-D equipped as a pullshovel for such work.

NORTHWEST

NORTHWEST

Two close-up views of the high mounting. The wider 36 inch track pads seen on this machine were available as an option. The crawler assembly was produced under a sub-contract with Allis-Chalmers, and later, Fiat-Allis.

Previous page: For the pipeline industry, Northwest offered a version of the Model 25-D with higher ground clearance and mounted on a tractor-type crawler, marketed as its "Hy-Trak" option. This unit is seen at the Northwest plant in Green Bay.

The Model 26

Produced from 1932 to 1939
1-1/2 cubic yard capacity

The rarest of Northwest machines: the Model 26 outfitted with even a rarer attachment of a large volume skimmer-scoop of 3 cubic yards. The Model 26 falls between the second and third generation machines. It carried the house lines of the former but the nomenclature of the latter. The machine was only offered with a gasoline power option. The operator's blindside still is of a squared off, box design.

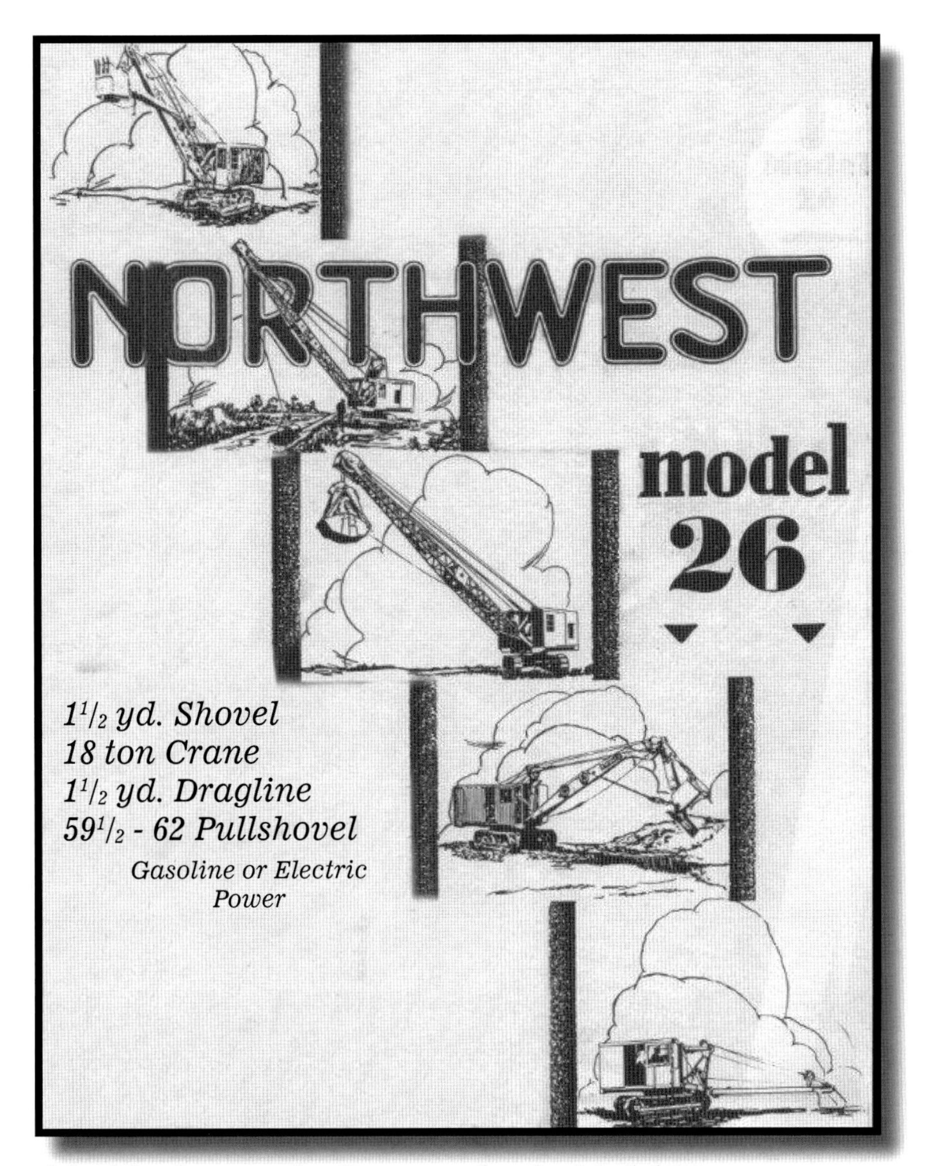

The front cover of the Northwest sales literature for the Model 26.

General Specifications Northwest Model 26	
Shovel Dipper Capacity	1½ Cu. Yd.
Hoisting Line Speed (Crane - Single Line)	150 F.P.M.
Dragline Cable Speed (Single Line)	123 F.P.M.
Standard Travel Speed	1 M.P.H.
Width of Crawler Treads	24"
Overall Width of Crawler Base	10' 0"
(L) Length of Crawler Base	13' 2"
(M) Height of Boom Hinge	5' 0"
(N) Height of Crawlers	3' 5"
(O) Clearance Under Rotating Base	3' 9"
(P) Overall Height of Cab	11' 9"
(Q) Rear End Swinging Radius	10' 7"
(R) Distance—Center Pin to Hinge Pin	3' 10"
Gasoline Engine	Wisconsin 6-Cyl.

The general specification chart to the Model 26, very little information is known about this rarest of Northwest machines. It was a machine of 1 ½ cubic yard capacity with similar dimensions to the Model 6. No factory records have been uncovered regarding tangible evidence for years of production or numbers of machines built.

The Model 40

Produced from 1931 to 1940
1 cubic yard capacity

This is one of the early Model 40 shovels seen at rest on a pavement removal job in front of the Fairmont Creamery on Broadway in Green Bay. These early machines had an "upward visibility extension" for the operator. Of particular note is the newly redesigned shovel attachment, with simplified crowd reeving, welded boom construction, and three-sided dipper bolted to a flared back which was part of the dipper stick. Note the sloped blindside of the machine a definite feature of a third generation machine.

The prototype Model 40 during an impact test in the factory yard at Green Bay. The photographs are a bit deceiving giving one the impression that the machine was driven off the end of perhaps a loading dock. The machine was actually jacked up and a rail flatbed car placed under the rear, and then pulled away to achieve the desired effect. We hope the operator didn't get a severe case of whiplash when the machine was suddenly dropped; no doubt the boom certainly did!

The Model 41

Produced from 1940 to 1943
Production interrupted by military production demands 1953 to 1980

Lots of action appears in this photograph on a cold December day in 1947. An early production Model 41 is seen loading out a Kenworth tractor / log trailer combination. As a Caterpillar RD-8 equipped with a Hyster winch skids fresh cut timber to the loading site. The truck driver walks a log and seems to be enjoying a smoke as he and his truck idly waits.

The same early production Model 41 as in the proceeding photograph is still working for Gordon A. MacGregor of Council, Idaho stacking logs with a straight steel heel boom. This image was dated January 6, 1949. This photograph and the previous page show a Model 41 with the original side window cab design. Take notice of the heavy pipe guarding on the cab.

Two "breadbox cab "Model 41 shovels featuring the later-style rollback operator's door.

Three later-year Model 41 pullshovels on various jobs performing a typical task: excavating a ditch for laying pipe. The Model 41 pullshovel boom was the only boom in which the bend occurred well beyond the center. These cabs now featured the rollback door design.

Two nice profile photographs depicting Model 41 draglines with the rollback-side cab at work. The bottom Model 41 utilizes a mast which was an uncommon feature for a Northwest dragline.

An elevated operator's station would enhance visibility in a number of applications, such as in a dockside operation, or timber stacking, or any other type of operation where the higher vantage point would prove beneficial. Northwest met the need with the Hi-Vue cab. A Model 41 crane at the Green Bay factory showing an early version of a pneumatically operated Hi-Vue cab. Once the cab was raised, the "skirt" below the cab with entrance ladder was installed.

A Model 41 with the factory installed Hi-Vue cab mounted on a 6x4 Maxi self-propelled wagon carrier. The early-style cab has been replaced with a new squarish design.

A few model 41's were outfitted on FWD 6x6 truck carriers. This rather unusual example was fitted with the standard pullshovel attachment for a Model 41, which would be the largest Northwest pullshovel to be mounted on a carrier. A bracket at the front holds the bucket while in transit; as an alternative, the attachment could be hung off the rear.

"But people, that is where the money is." This quote from President of Northwest Everett Windahl was in response to a complaint offered up by production management when special application machine orders were cutting into the main production schedules. Northwest would generate large profits from augmenting regular production models with special purpose units. This special application Model 41 shovel tunnel mucker, with a short boom, is a typical example of such reworking to fit the customer's needs.

Two photographs taken at the factory in Green Bay of the 60 ton Model 41-T truck carrier mounted crane. The crane in the top image shows a unit undergoing testing for longtime Northwest customer Herbert F. Darling Engineering Contractors of Buffalo New York. The bottom view shows the operator's blindside and extended mechanical outriggers along with a nice close-up of the Dart 8x4 truck carrier.

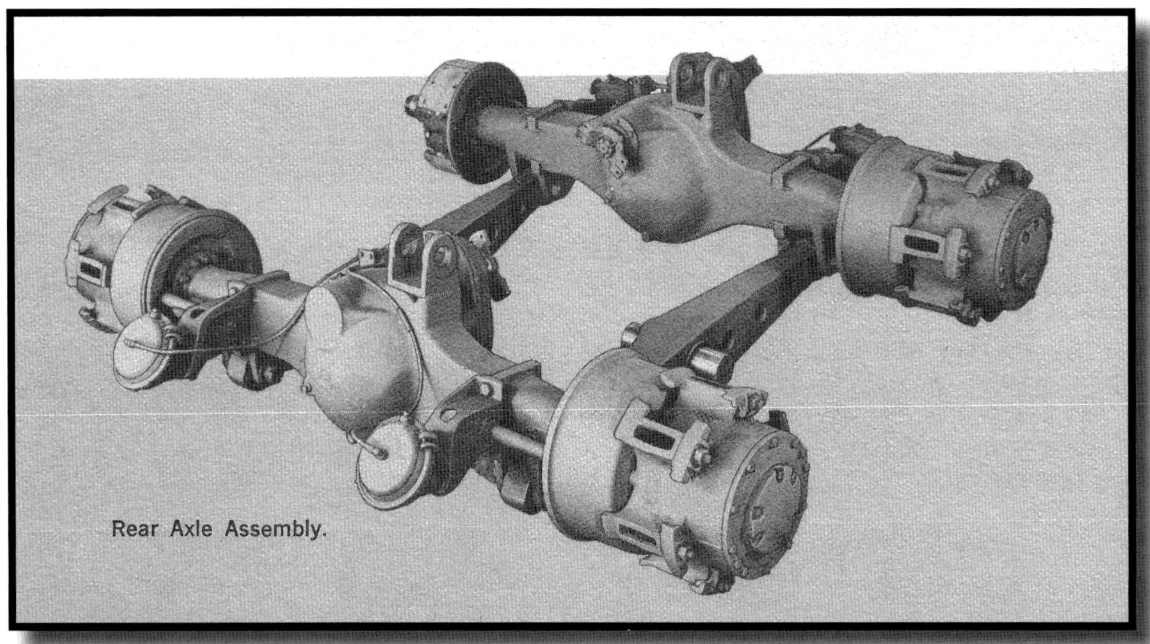

Here are three illustrations of an F.W.D. 8x4 truck carrier and its various components. F.W.D. Corporation of Clintonville, Wisconsin, was the primary supplier of truck carriers for Northwest. A major feature of a F.W.D. carrier was its box shaped axle construction as shown in the center illustration. This was a carryover from their heavy-duty truck designs.

Parked in front of Northwest dealer Cal-Ore Machinery Company of Portland, Oregon, is a bare-bones Model 41 Timber-Master, without boom, on its Pierce Pacific 8X4 truck carrier prior to delivery from Northwest Engineering to customer Wesley Graves. Cal-Ore would install the customer required arch heel boom and cable operated logging tong assembly.

This and the following page show the newly delivered truck carrier mounted Model 41 TM (Timber-Master) at work in its intended environment. Fitted with the arched heel boom and grapple, it was an efficient log-loader in the back country. Notice the hydraulically operated outriggers.

WESLEY GRAVES

WESLEY GRAVES

An excellent detailed view of the Wesley Graves Model 41 Timber master with rugged heel boom and its mechanical cable operated logging grapple.

Not all Northwest's were mobile, quite a few were supplied through the years as a pedestal-mounted unit. Such was this Model 41 outfitted with a crane and magnet for a scrapyard application for Industrial Metals of Toronto, Canada.

The Model 6 updates

The Model 6 received all the house sheet-metal updates as with the rest of the Northwest line. Here a Model 6 shovel sporting the classic "breadbox" house loads a Mack AC dump on a road renovation job for Brautigam Trucco Company.

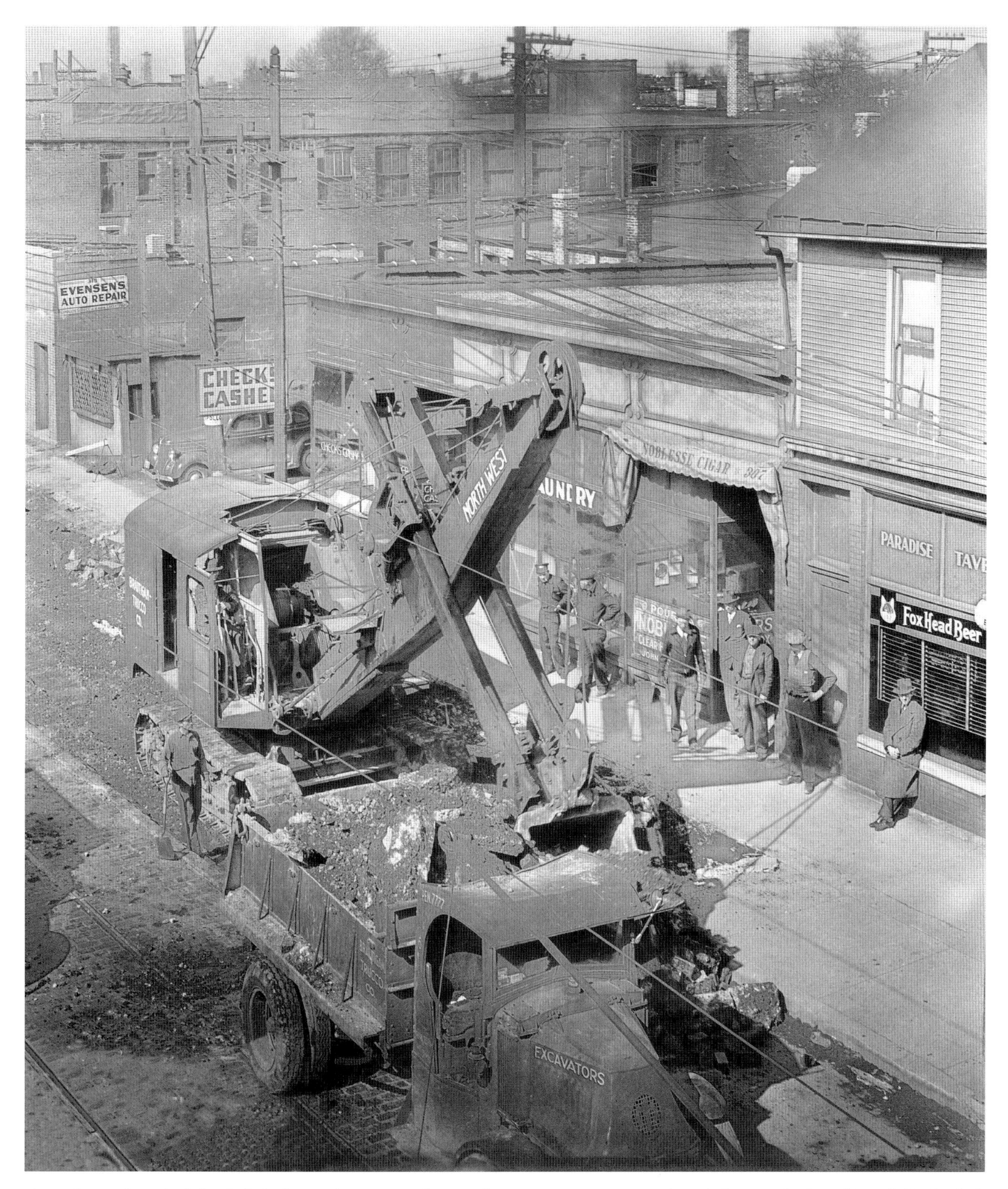

Another classic Model 6 shovel on another urban street renovation job, again loading Mack AC "Bulldogs". The gentlemen seem to be enjoying the activity. Before front-end loaders came on the scene, shovels were used for pavement removal.

An early 1950s Northwest catalog cover to the Model 6 shovel with a side window cab. The shovel seems to dwarf the truck being loaded. Model 6 shovels did not have gantries, and utilized the twin-stick shovel front.

A mid-1960s Model 6 shovel with a rollback door cab loads out a Haul-Pak truck with shot rock.

Two Model 6 pullshovels updated with the "breadbox" style house with gooseneck booms: the one in the top photograph is working in Massachusetts, outfitted with the older riveted boom and dipper stick. The bottom photograph shows the newer welded boom and stick and redesigned bucket. The Model 6 in the lower photograph is working in its hometown of Green Bay on a residential sewer job for Chapel & Amundsen.

Two more Model 6 pullshovels are shown on this page. The photo above shows the machine with the older riveted gooseneck boom but with a welded stick and the older bucket design. Also note the low mounted rear gantry. The unit is working on a sewer job in the Kansas countryside lifting concrete pipe in preparation for placement. The Model 6 pullshovel shown to the left has a revised boom and bucket, and is working on a sewer job, for Verona Construction Company of Verona, New Jersey. A portion of this rocky job was right through the middle of a marina hence the presence of the boats parked alongside. This Model 6 with the new boom and bucket has a cleaner look.

Setting reinforced concrete pipe down into the ditch is the task at the moment for this Model 6 pullshovel, equipped with the new-style boom and bucket, before resuming opening the trench. Skillful handling of the brakes was required for precise placement of the pipe. The digging appears to be in clay, judging by the bucket teeth marks in the material excavated. There is a lack of shoring in this deep excavation.

Casting for pulpwood". Model 6 cranes unload out bundles of pulpwood on the St. Lawrence River for B.F.D. Paper mills located in Ogdensburg, New York.

Another view of the unloading operation with Model 6 cranes.

A new Model 6 crane mounted on a 6x4 Maxi wagon carrier for Juno Prestressors Inc. is being readied for delivery in May of 1960 at the Green Bay factory.

Another Model 6 mounted a Maxi 6x4 wagon carrier doing drop ball work in a stone quarry operation.

A Model 6 dragline with a forward A-frame gantry loads out Mack LT six-wheel dump trucks in a sand and gravel operation in the mid-west United States.

Another "classic years" Model 6 dragline.

A Model 6 equipped as a crane outfitted as an air operated pile driver with a Vulcan single-acting hammer. This Model 6 has the riveted rearward gantry. It appears to be working in a railyard.

A late year rollback side Model 6 crane is setting piles for concrete bridge piers with hanging leads and diesel hammer.

A very unique Model 6 combination crane on a tram sled doing stockpile work off a pier.

The Model 70

Produced from 1933 to 1941

A Model 70 shovel equipped with a fairly wide undercarriage pioneering in what looks like timber country. Notice the upward visibility extension on the cab.

(Item of note: The Model 70 of this era was a model line branded with this nomenclature. The Model 70 line introduced in the late 1960s was an entirely different machine developed along different design lines. The only thing the two machine lines shared was the common name of their manufacturer, Northwest.)

Another Model 70 shovel resets timber mats for the operation. Could those barrels in the foreground be filled with whiskey or bourbon for the men? Most likely they are filled with fuel for the machine. Northwests found ready acceptance in big timber country.

The Model 71

Produced from 1937 to 1943

An early Model 71 erection crane is seen working here for Bethlehem Steel. Note the light jib atop the boom.

A Model 71 crane working in Hawaii at Pearl Harbor for the United States Navy. The Model 71 was designed from the ground up as a heavy lifting unit. No provisions were made to have the unit convert over to a shovel or a dragline. Special attention was given to the boom engineering and design. The boom metallurgy was of a special high silicon alloy steel that gave tremendous strength to weight ratio for its day. A high speed boom hoist worm gear drive with separate direction control was also a major feature along with a tall rear mounted A-frame gantry.

Another Model 71 crane set up as a piledriver is seen setting piles from a barge on the Rhine River in Germany for the United States Army Engineers Corps after the end of hostilities of World War Two in the European theater of operations. Many Rhine River bridges had to be rebuilt by the occupying forces since the German Wehrmacht had orders to destroy all river crossings to hinder the Allied advance during the war. This crane with hanging leads features a Vulcan hammer powered by the air compressor seen just to the right of the crane.

The Model 78- D

Produced from 1931 to 1943

Two Model 78-D shovels from the first batch produced. Notice their overall stoutness.

The cover to a 1936 sales literature catalog featuring the Model 78 2 cubic yard shovel. This particularly handsome cover, with a beautiful profile rendering of the machine, was typical of the beautifully illustrated literature promoting the features of Northwest machines, produced by Russell T. Gray, an advertisement agency in Chicago.

A nice profile photograph from the same 1936 Northwest sales literature.

A Model 78-D handles a rather large boulder. Upgraded features were welded cambered boom and the attachment of a three-sided dipper to the flared back which was part of the twin dipper sticks. These upgrades gave great strength to the shovel attachment for heavy rock work.

A Model 78 shovel loads out a Caterpillar D-8 pulling an Athey crawler dump wagon. The Model 78 was Northwest's big boy for digging for many years. At 2 cubic yards it would be the pinnacle in shovel design until the advent of its bigger brother, the Model 80. The 78-D was a morphed variation of machines utilizing a travel base of a Model 8 and a house and draw works from a Model 6. The boom and dipper was a new up-engineered design having all the same design criteria as the smaller 1- ½ cubic yard Model 6.

The Model 80 and the Majestic Model 80-D

Produced from 1933 to 1973
The early square box style cab and latter classic "breadbox" style cab

NORTHWEST'S Living Legend

1933. Franklin D. Roosevelt began his first Presidential term, Prohibition was repealed, John Dillinger was terrorizing the Midwest, and Northwest Engineering brought out its first Model 80. The basic machine with a 2-cubic yard shovel front carried a list price of $23,850, F.O.B. Green Bay, Wisconsin, then as now NORTHWEST's only plant location.

Since then, the U.S. has had seven Presidents, Prohibition is a dim memory and Dillinger, killed by the FBI in 1934, became a legend.

So did NORTHWEST's 80, but it's a living legend, still being built and going strong. Although early records are vague, an estimated 2,600 80's were shipped in the past 48 years. Company officials believe parts sales show at least 1,500 80's working worldwide as shovels, pullshovels (NORTHWEST's name for its backhoe) and cranes.

Many contractors swear by the basically unchanged machine, claiming that, because of lower maintenance and less down-time, it turns in lower production costs than a wheel loader.

Subtle Changes

Although larger machines - many hydraulic - and wheel loaders dominate the market, probably no excavating machine in its class is as indelibly etched in the memory of the men who move rock and dirt for money as the 80-D workhorse (No one at NORTHWEST today remembers why the "D" was added in 1937).

The rig has grown in power, capacity, weight and price, but its appearance has changed only slightly. The NORTHWEST profile is unmistakable, and design changes to the boom, track frames and cab have only barely altered the 80-D's appearance. Despite dipper stick redesign to a single stick in 1955, NORTHWEST machines have always used a cable crowd.

The earliest 80's were powered by four-cylinder Twin City gasoline engines with an 8-inch bore and 9-inch stroke and developed 120 HP at 565 RPM. Murphy diesel engines were installed in 1938. Twin City modified combustion, and for a few years NORTHWEST advertised an optional "oil engine that is easily understood by the gas-engine mechanic." Hand cranked, the magneto-equipped oil engine started on gasoline and switched over to "ordinary furnace fuel" after running up to speed and temperature.

By 1935, NORTHWEST offered six-cylinder Atlas or five-cylinder Fairbanks diesel engines for $4,500 extra, but even after Murphy replaced them, the gasoline-powered Twin City was very much a part of 80-D production until 1941.

For many years the factory shipped machines without dippers to owners and dealers, who then installed 3-cubic yard dippers in the field, theoretically voiding the 2½-cubic yard warranty. That changed in 1974, when NORTHWEST officially adopted the larger size dipper.

The original 80 succeeded NORTHWEST's Model 8, a shovel and backhoe that got its 2-cubic yard designation from its dipper capacity in numbers of ¼ cubic yard. 131,000 lbs. weight in 1933, current 80-D shovels weigh 149,000 lbs. Price hikes have been more precipitous: Today's 80-D lists for $367,725.

Explaining the Mystique

John Civetta & Sons, Inc., a New York City heavy contractor, bought its first 80-D in 1954. They now own five. "They're bought and paid for," says company treasurer, John Civetta, Jr., "and for most of our work they're as good as a new machine. We can run one of them on 40 gallons of fuel oil a shift if the going isn't too bad. A (medium class) crawler loader burns 60-70 gallons every eight hours. In Manhattan, with heavy traffic to hamper trucks and close working quarters, we can't often get into high-production work, so a 2½-cubic yard cable hoe gets me as many yards a day as a 5-cubic yard tractor."

All Civetta NORTHWESTS today are backhoes or cranes. Charles Santoro is now dredging out a pond in Manhattan's Central Park with a 10-year-old dragline with a 60-ft. boom and a 2½ yard perforated bucket. When he started operating, right after World War II, "almost everything was manual, and crater compound used to leak so badly from the bevel gear sets that the oiler would have to keep throwing fuller's earth — sometimes loose cement — into the frictions to keep the brakes dry."

"Times have changed," Santoro continues, "but I'm one of the few guys left who would rather run a cable hoe than a hydraulic machine. Some of the quarries are going back to shovels because they don't wear out tires or blow hydraulic hoses. They're just plain cheaper to operate."

And so the Living Legend, the NORTHWEST 80-D, lives on . . . and on . . . and on.

10

An article penned by Northwest and placed in the second to last issue of *Material Handling Illustrated* dated 2nd quarter of 1981 pertaining to its legendary Model 80-D. John Civetta is a prominent construction company in the New York City area that was a steady customer of Northwest machines.

A Model 80 dragline employed by John Marsch Inc. sets a new grade for a rail bed relocation project on the Wabash Railroad near Chicago, Illinois. The early Model 80 used what was known as a rearward gantry, so named because it pointed toward the rear of the machine. This particular example shows the extension that could be added above the gantry to raise the boom hoist cables higher to reduce boom compression. Also take notice of the square box style of the machinery house with the upward visibility "raised eyebrow" extension for the operator.

A Model 80 crane equipped with a pile-driver working for Mecca Construction of Clearwater, California in March of 1938. This machine seems unusual in that it appears that the hammer is operating off a steam boiler seemingly mounted within the machinery house of the crane.

In September of 1939, a Model 80-D dragline loads out a Euclid tracked dump wagon being pulled by a Caterpillar RD-8. All units are in the employ of Ralph Myers Construction of Salem, Indiana.

A Model 80 equipped with a field improvised twin timber boom loads out big timber for the McDonald Logging Company of Jackson, California. One may surmise that this machine may originally have been supplied with a shovel front-end, hence the rearward gantry, less the crane extension more often used on this style gantry for an operation like this. The scene took place on March 10, 1942.

A Model 80 works off of a timber-walled river bank setting a large diameter section of steel pipe for Utah Construction in May of 1949.

The very first Northwest Model 80 shovel, this photo dated July 2, 1933 shows the company's hallmark machine with its most recognizable attachment at its debut rollout at the factory.

Another of Northwest's Model 80 outfitted as a shovel loaded on a rail flatcar and ready for delivery from the Green Bay factory to a customer in Oregon on April 13, 1938. This machine with this attachment, more so than any other Northwest product, would be the pinnacle to which other full revolving friction-based machine manufacturers would have to aspire to. Over the Model 80-D career, Northwest would ship an estimated 2,600 of the proven performers in all of its configurations.

A Morrison-Knudsen Northwest Model 80-D shovel early model loads out a string of side-dump railcars near Boise, Idaho in May of 1939. The 'D' suffix was added in 1937, changing the model designation from the 80 to the 80-D. This one still sports the early square box house.

An early Model 80-D shovel cuts a mountain notch roadbed and loads out a caravan of Allis-Chalmers crawler tractors pulling what looks like Euclid track dump wagons.

The power for any grade, with a little help from two Caterpillar D-8 2U crawler tractors, that is. Utah's Construction's Model 80-D crane needs just a little bit more to make the steep climb up what appears to be a 45 degree slope. We hope the company's mechanic had recently inspected the machines brakes!

This Model 80-D dragline working in July of 1947 for B.I. Construction of Pittston, Pennsylvania is one of the first units shipped with newly redesigned house for a Model 80-D. The famous "breadbox" style house with its rounded roof lines would be one of the most identifiable features to a Northwest for years to come. This handsome Model 80-D dragline frames a portion of the picturesque Susquehanna River community of Pittston.

Brunswick Pulp & Paper Company utilizes a Model 80-D crane equipped with an orange peel grapple to sling logs at their mill in New Augusta, Georgia. Another Model 80-D works the opposite log windrow in the background and further back a Model 95 with an additional external counterweight handles even more logs.

Following page: Iskra Brothers Logging Companies of Aberdeen, Washington Model 80-D handles big timber with a heel boom and a log tong without any issues.

Kosmos Timber of Kosmos, Washington employs a heel boom, tong equipped, rearward gantry Model 80-D on one of their clear cut operations in the late 1940s.

A close up of the above machine showing the details for the reeving of the hoist cables and construction of a heel boom.

Mount Whitney Lumber located in Johnsdale, California uses a rearward gantry Model 80-D twin timber boom machine. Note the sectional construction of the boom with the insert feet, tube splicing coupling mid sections, and boom head.

Two different Model 80-D's with two different style heel booms. The upper Model 80-D has the older taller forward mounted A-frame gantry and heel boom; while the lower is a newer machine with the lower rear mounted A-frame style and a heel boom manufactured by Albin Manufacturing and a Ritz log grapple.

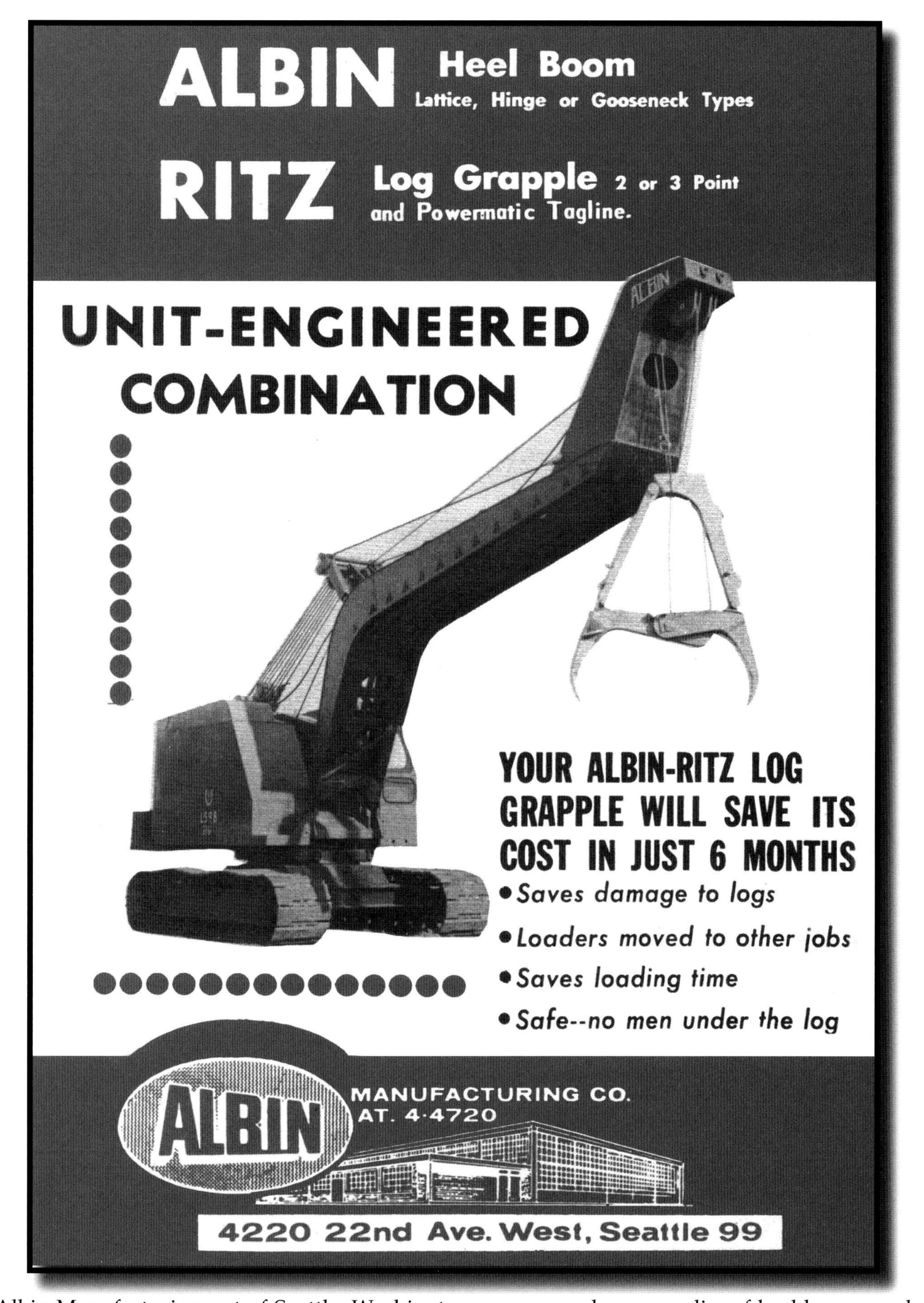

Albin Manufacturing out of Seattle, Washington was a very large supplier of heel booms and associated logging attachments, not only for Northwest machines, but other full revolving machine manufacturers in the United States. Note the above machine is a Link-Belt LS-98.

The first Model 80-D shovel with the redesigned "breadbox" cab dated May 1, 1947. The machine still retains the older style twin dipper stick front-end along with the rearward gantry.

An 80-D shovel loads railcars powered with a steam locomotive.

This and the preceding page show one of the first Model 80-D shovels with the redesigned "breadbox" house working in North Carolina on the Atlantic Coast Railroad in June 1947.

Following page

Top picture: Freeport Sulphur Company's Model 80-D shovel, the machine appears to be equipped with a much narrower undercarriage than a standard issued unit.

Bottom picture: Another Model 80-D shovel loads out a string of side dump rail cars for the Matt McDougall Company of Portland, Oregon in July of 1949.

NORTH WEST
NATT McDOUGALL
COMPANY
CONTRACTORS

This Northwest advertisement featuring Model 80-D machines that heavy highway contractor Dwight W. Winkelman of Syracuse, NY would purchase, along with the advertisement shown on the next page highlights customer acceptance for Northwest products, and repeat orders depict Northwest machines as a money maker for many prominent contractors.

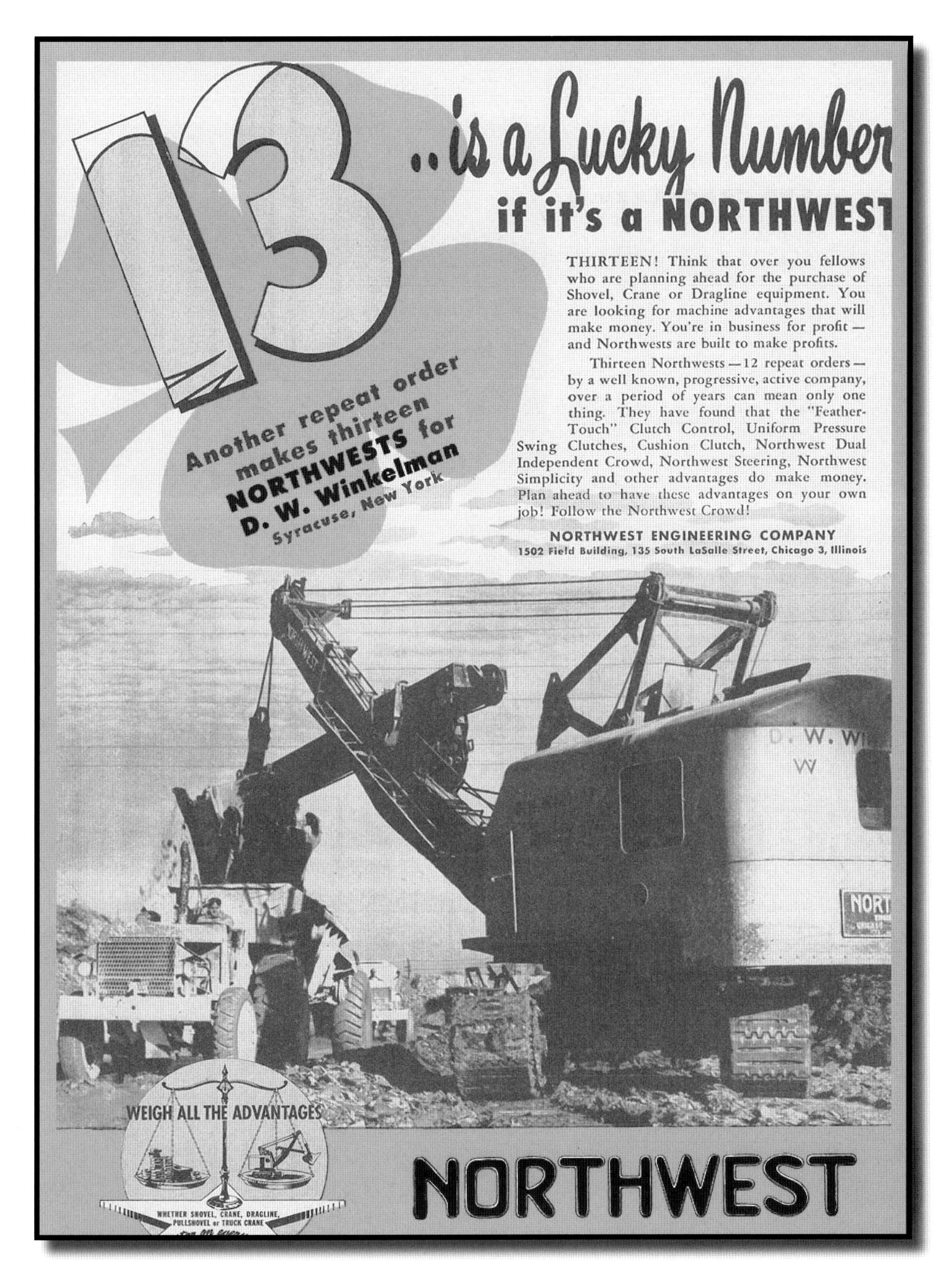

Another advertisement highlighting repeat orders by Dwight Winkelman. Apparently, the number thirteen was a good one for the company for the total number of Northwest machines ordered.

An 80-D Pullshovel digs with its bucket fully extended at full-reach on a foundation job in Chicago. In the Northwest sales literature, Northwest emphasized the long reach of its pullshovel. Take notice of the flared sidecutters (elephant ears!) on the bucket. Also absent in the background are barricades, permitting onlookers to enjoy the digging action. Chicago was the sales headquarters for Northwest.

Another view of the same pullshovel loading soil into what appears to be a Mack LJ dump truck with a rock body. The unmistakable profile of the Northwest 80-D pullshovel, with its distinctive house design side window cab with stout shear-leg, cambered boom, and stick with swiveling sheaves at the top and unique bucket are all evident in this view. There would be many construction sites throughout the United States during the 1950s that had a Northwest Model 80-D pullshovel digging proving its worth. The minimal downtime to a Northwest was strong selling point to many customers.

Two later day roll back door Model 80-D pullshovels, the upper unit is excavating a building site on a heavy highway project in New York City, and the lower unit excavates a riverbank for a new bridge.

The 4th quarter 1966 *Material Handling Illustrated* back-cover advertisement for a Northwest Model 80-D pullshovel equipped with a Hi-Vue cab option.

How to drill grid holes on a 30 ft. rock face was a problem on a Kansas City freeway. Anchor bolts were specified on 6 ft. centers. Clarkson Construction Co., solved the problem by hoisting a track drill with an 80-D Northwest. This proved safer and faster than alternately lowering a man with a jackhammer. Platforms were provided for the drill operator and his assistant.

It isn't often you need a jumbo in a hurry. Clarkson Construction Co. of Kansas City on one of their jobs had grid holes to dig in a 30 ft. rock base. So they hoisted drill rig and crew on their 80-D Northwest. The shovel moved along for the range of holes and the job was faster than it could have been done by lowering a man and jackhammer.

Really, what would OSHA say? The above two photographs and captions were taken from of two separate Material Handling Illustrated issues, dated respectively 1966 and 1967, both of the 4th quarters, showing the use of a Northwest 80-D as a makeshift drill jumbo.

Two photographs of the first Model 80-D with the newly designed front shovel attachment featuring a single dipper stick and split boom. The left photograph was taken during the winter of 1952 when the unit was still undergoing engineering development machine minus the dipper and padlock has the new rear mounted A-frame gantry. The device mounted to the shovel boom foot is a jig designed to test the side stain imposed when digging. The right view was taken in the summer of 1954 after the unit was received back at the Northwest plant following its field testing. The revised Model 80-D with its new modern front-end was released for sale in the spring of 1955. This by far would be the most recognizable variant of Northwest's most machine ever produced.

The famous Northwest "Breadbox "

This is the back and operator's side views to a Northwest "breadbox" revolving house. Through the square opening, the legendary Murphy Diesel MP-21 can be seen. The mating of the Murphy Diesel and an 80-D was an unbeatable match. Item of note, Northwest would offer as a complimentary service the lettering of the new machine owner on the side of the house. This was a practice that was exercised by the company its entire 70 year history.

The first "roll-back-side" house Model 80-D shovel. The door panels slide back about half the length of the machinery house. It is shown here performing its swing tests on May 6, 1959 near the edge of the Fox River, with the city of Green Bay in the background. Other changes visible in the photograph, is the A-frame gantry, which replaced the earlier "rearward" gantry, along with a newer simplified split boom with single dipper stick. These changes created the definitive version of the 80-D shovel. Compare this profile of the 80-D shovel with pictures of earlier versions of this machine. All versions of the 80-D shovel were hard-working, reliable machines designed for tough digging conditions, earning them the moniker "a real Rock Shovel." Of particular interest on this machine, is the door which has been slid back behind a swinging panel.

As the ad says, "a real rock shovel doesn't just happen". No more really has to be said. Compare the door arrangement on this machine and the one on the previous page. Here the operator's door slides back over another panel which also slides back. This was the style used on all later rollback-side machines.

Two Model 80-D shovels with rollback-side, breadbox style cabs in action. These machines still retain the mechanical Feather-Touch controls. Many a Model 80-D shovel found a lifetime home in a rock quarry operation. The lower machine has early door arrangement.

Thailand

Formosa

The Canary Islands

Above is a series of photographs that appeared over the years in Northwest's *Material Handling Illustrated* showing various Model 80-D shovels working overseas, all working in rocky conditions.

An overseas modified Model 80-D twin-stick shovel is seen loading trucks (or as the Aussies say, "lorries") with basalt in Tasmania, Australia. This 80-D was originally supplied to the United States Navy and was a veteran of World War II, and then it was purchased by the Hobart Quarry operations. For unknown reasons, this machine's revolving house has been highly revised with a stepped boxy raised roof and remodeled operators cab.

And how they got there. Model 80-D's being loaded on a freighter bound for Iran; the only way for a Northwest machine to reach an overseas customer was via ocean transport. The top photo shows a newly assembled Model 80-D shovel loaded and ready to be shipped via rail to perhaps a coastal seaport. Until the advent of the St. Lawrence Seaway in 1959, Northwest machines would be shipped by rail to a port and then loaded onto freighters bound for foreign customers. After 1959, the docks on the Fox River in Green Bay could be utilized to load out foreign orders depending on the ship's draft. Many machines were shipped either by truck or rail to the port of Sturgeon Bay, approximately 30 miles north of Green Bay where the water's depth could accommodate larger vessels.

A sequence of photographs showing the loadout of a new Model 80-D shovel bound for Sweden. This Model 80-D was loaded aboard the S.S. Monico Smith registered with the Swedish-Chicago Lines. The Model 80-D was moved from the cab assembly building via a lowboy provided by Leicht's Transfer & Trucking of Green Bay to the Northwest factory dockside on May 7, 1959. This was the first international shipment for a Northwest machine by freighter through the Great Lakes via the newly opened St. Lawrence Seaway to the Atlantic Ocean.

The Model 85

Produced from 1933 to 1936
A prelude to the Model 95

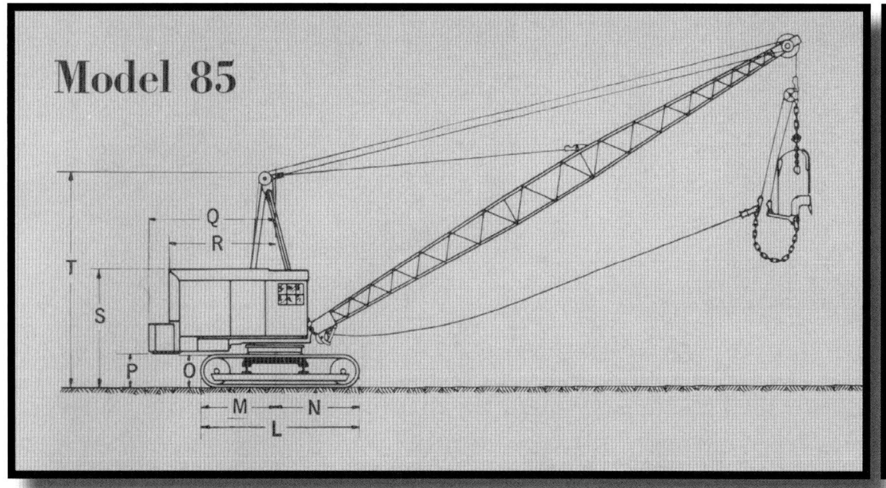

General Specifications

Northwest Oil Engine—4-cylinder	8″ x 9″
Twin City Gasoline Engine—4-cylinder	8″ x 9″
Diesel Engine or Electric Motor	On Application
Width of Crawler Treads	33″
Overall Width of Crawler Base	12′ 3″
(L) Overall Length of Crawler Base	17′ 5″
(M) Length of Rear Half of Crawler Base	8′ 2″
(N) Length of Front Half of Crawler Base	9′ 3″
(O) Height of Crawlers	3′ 5″
(P) Clearance Under Rotating Base	
*Including Fuel Tank on Rear End	3′ 8½″
Not Including Fuel Tank on Rear End	3′ 11″
(Q) Rear End Swinging Radius	
*Including Fuel Tank on Rear End	13′ 9″
(R) Rear End Swinging Radius	
Not Including Fuel Tank on Rear End	11′ 7″
(S) Overall Height to Top of Cab	12′ 6½″
(T) Overall Height to Top of Gantry	22′ 9″

*Fuel Tank on Rear End is standard on Model 85 Dragline *only*.

BUCKET CAPACITY DEPENDS ON WORKING RADIUS REQUIRED, LENGTH OF BOOM AND THE NATURE OF MATERIAL TO BE DUG OR HANDLED.

The Northwest Model 85

This is the cover for a catalog produced for the Model 85. The Model 85 was only offered as a dragline. The 85 was initially introduced at 2-¼ cubic yard capacity, but soon was upgraded to 3 cubic yard capacity at customer requests. There is no known factory production numbers for the Model 85.

An Extra ¼ yd. Capacity

with Bucket Life that Has Never Been Equalled!

The Northwest Bucket has not only made greater yardage possible with your dragline, but it brings you the quality and strength that only high grade steels can give. It assures you extra yardage day after day and month after month without constant rebuilding. Compare it with your present bucket!

- The arch—a heavy I-beam section, not to be compared with the usual plate construction that must be regularly reshaped.
- The lip—rolled alloy steel, heat treated for extra toughness.
- The basket—formed and welded by a special welding process to become one piece.
- Hoist trunnion pins of hardened steel.
- All castings heat treated for extra toughness!

See them in operation. Compare their output with what you are getting—talk to users.

Ad copy used in Northwest sales literature touting the advantages of their 3 cubic yard capacity dragline bucket. It is not known if these buckets were made in-house or sourced from an outside vendor.

Northwest developed the Model 85 specifically for the surface coal mining industry. While there were certainly other manufacturers that had larger capacity machines at the time that catered to this industry, Northwest wanted to tap into the smaller surface operations that dotted the landscape throughout the coal regions. The Model 85, with its 3 yard capacity, fit this niche market very well. The Model 85 would carry on the legacy that Paul Burke started with the Type M dragline back in 1923. This legacy would eventually develop into one of Northwest's all-time great dragline equipped machines, the Model 95. Note the exceptionally tall forward gantry helping with reducing boom compression, enabling a longer boom and larger bucket. The majority of Model 85's were shipped with Northwest's Oil engines enabling long working hours at low operation costs.

The Model 95

Produced from 1936 to 1973
The early square box style cab and latter classic "breadbox" style cab

The cover of the 1938 Northwest sales literature catalog for the Model 95. Introduced in 1936, it featured longer crawlers, which made it ideal for dragline use. This early model had a high, forward mounted A-frame gantry.

Following page:

Top picture: Two Model 95 draglines remove overburden while a Model 80-D shovel works the exposed coal seam. Note the Model 80-D was originally shipped with a long boom hence the extended rearward gantry.

Bottom picture: An early Model 95 dragline loads the Scandia's Gold mine dredge in House Creek, California in October 1941. Similar to the Model 80, it had a box-style machinery house with the upward visibility extension for the operator.

SCANDIA MINE
ROBINSON MFG. CO.

NORTH WEST
DRAVO

A Model 95 with another field improvised timber boom, this particular unit has heavy steel plate bolted to the bottom side of the boom to use it as a heel boom. The unit is seen working in Longview, Washington in 1949 for Weyerhauser Timber Company. This image was used for many years in Northwest sales literature for the Model 95. This unit has a rounder, newer style machinery house with a side window by the operator; note the tall forward A-frame gantry of earlier models.

Previous page: Two early Model 95 machines in different applications, the upper photograph showing a dragline working for Dravo, and the lower photograph showing a log loading operation. Both are of the original Model 95 types with the tall forward gantries and "raised eyebrow" vision cabs.

The cover of the 1957 sales catalog for the Northwest Model 95, these handsomely designed pieces of literature were produced by longtime Northwest advertising agency Russell T. Gray Company of Chicago, Illinois. In the 1950s, the machinery house of the Model 95 acquired the distinctive, recognizable "breadbox" style, with rounded contours. This model features the "side-window" cab. Also take notice of the redesigned, rear-mounted A-frame gantry. The longer house on a Model 95 contained a 300 gallon fuel tank, as compared to the shorter house on an 80-D.

One of Northwest Engineering Company's many prides, a Model 95 dragline. With the "bread-box" style house and rear-mounted gantry, a Northwest machine cut a unique profile.

A Model 95 pullshovel at work. It has the old gooseneck boom, rearward gantry and a short shear-leg (mast). This was an interim machine with the updated side-window "breadbox" machinery house. Another open trench without shoring is shown, which would give fits to OSHA today.

Another interim Model 95 pullshovel at work.

Dig deep! A Northwest 95 pullshovel does just that on this job. The deep ditch is being taken as a double-cut, with men way down in the hole without the protection of a trench shield. Another OSHA no-no!!

Two later year Model 95 pullshovels performing trench work. Again, OSHA would have a field day with the working situations depicted in the above photographs.

Two views of classic Model 95 pullshovels.

Two operator side views of rollback door Model 95 pullshovels The lower photograph shows a machine with Cushion-Air controls operated by air, replacing the former mechanical Feather-Touch controls in 1965. Cushion-Air controls first appeared on the larger 180-D machine introduced earlier in 1962.

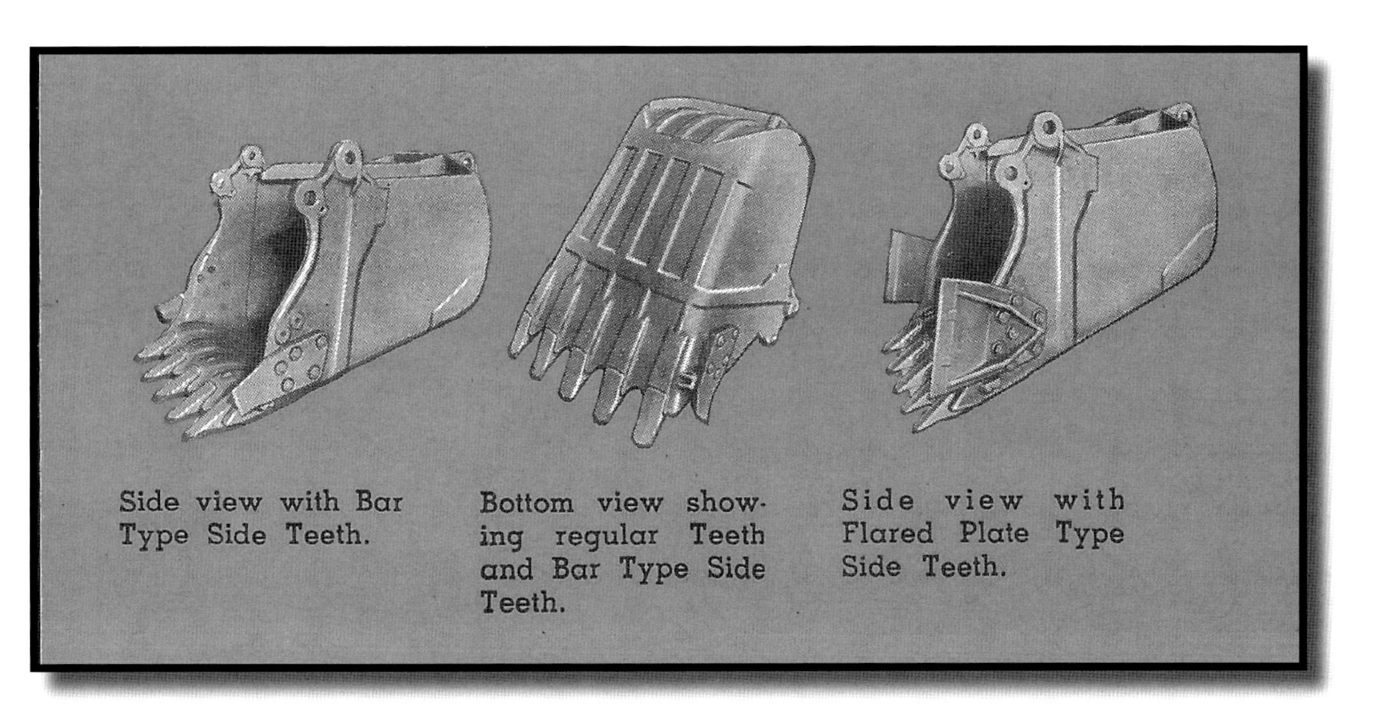

Side view with Bar Type Side Teeth.

Bottom view showing regular Teeth and Bar Type Side Teeth.

Side view with Flared Plate Type Side Teeth.

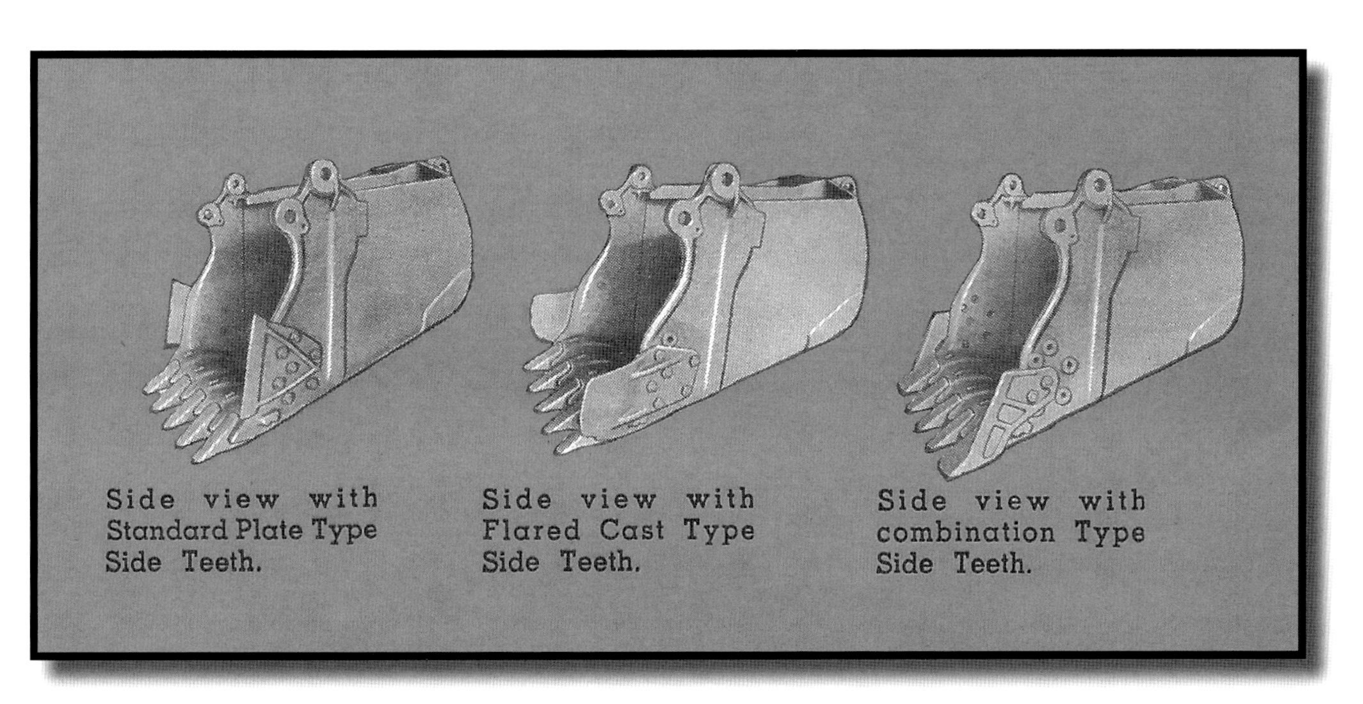

Side view with Standard Plate Type Side Teeth.

Side view with Flared Cast Type Side Teeth.

Side view with combination Type Side Teeth.

Pullshovel buckets shown with different styles of side-cutters depending on the digging conditions were offered. They were made with a heavy-duty cast front section and lip welded to a fabricated body.

A brand new Model 95 dragline minus its bucket swings a weighted load tethered by the machine's drag cable for its initial run-through testing prior to delivery to a customer. The machinery house has undergone yet another upgrade to a "rollback-side" type, similar to the upgraded 80-D shovel shown previously. The longer house contains the larger 300 gallon fuel tank, and distinguishes the Model 95 from the shorter 80-D, although the internal draw works are the same. As shown earlier, another upgrade has taken place with the gantry, which is now a rear-mounted A-frame type. A portion of the city of Green Bay is seen on the other side of the Fox River on March 7, 1959.

The same style Model 95 set up as a dragline. This machine does not feature a boom hoist bridle, as in the top photo, but rather is reeved continuously from the gantry to the boom head point.

The Model 95 was also a much desired machine for general crane lift work.

A Model 95 crane is seen here setting sheet-piles.

Two new Northwests working on an interstate overpass, a Model 95 with a jib sets an I-beam with a little help from a smaller brother Model 41 equipped with an additional external counterweight. Northwest machines contributed mightily to the Interstate construction program.

A Model 95 crane rated at 60 tons fitted to a 6x6 Maxi 6460 wagon carrier with hydraulic outriggers on November 18, 1965. This would be the largest combination machine/wagon carrier offered by Northwest. This unit has the fully upgraded machinery house and features a new method of control from the earlier mechanical "Feather-Touch" clutch control to air controls, known as "Cushion-Air".

This Model 95 Mobile-Master is handling pre-stressed concrete bridge beams for Juno Prestressors of West Palm Beach, Florida. Juno was a very large customer for Northwest for their wagon crane line which was billed as Mobile Masters. This photograph appeared in the 2nd quarter 1966 issue of Northwest's *Material Handling Illustrated*.

The Model 180-D

Produced from 1962 to 1969 (Series I)

A Northwest advertisement image showing the digging dimensions of a Model 180-D. The 180-D was a substantial leap in size of machine produced by the company.

A view of the operator's blindside of the prototype Model 180-D shovel, September 1961. The machine is still in its primer brown guise.

The prototype Model 180-D shovel just after painting the house in Pekin orange and gloss black for the shovel attachment and undercarriage, but it appears to be without its black roof trim.

A new form of air controls for Northwest machines would begin to replace the tried and true Feather-Touch mechanical controls. The control console with short-throw levers shown here would replace the long-throw levers of the mechanical era.

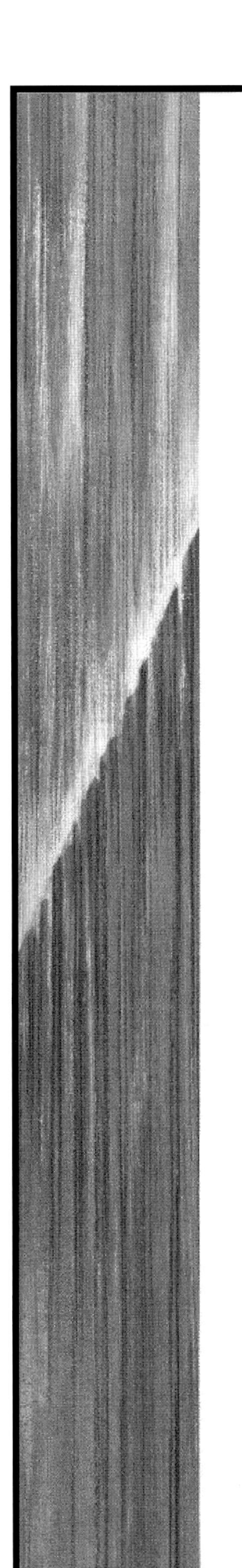

GIVING THE OPERATOR A BREAK!

A man at ease means more output!

Few operators of shovel and crane equipment will remember the days when shovel men stood to operate their rigs. It was Northwest that first gave throught to the improvement of shovel and crane control and greater operator comfort as an important influence for increased output. In 1924 Northwest brought to the operator the first real ease in gas machine operation with the introduction of the Feather-Touch Control. This was the first step in giving the operator a chance to better his output and make his job even more important.

The Feather-Touch Clutch Control utilized the power of the engine to throw the heavy drum clutches. It brought the feel of the load, does not drag on the drum, and the action is in direct ratio to the action of the operator's lever. At the same time, and for the first time in shovel and crane history, the operator found himself sitting down. The Feather-Touch Control is still standard on the Northwest 25-D and it is the finest, most responsive manual control ever developed.

In answer to the better development of larger machines and larger clutches, and considering the fact that an operator is expected to produce, Northwest developed an advanced design of air controls.

Northwest Cushion-Air Controls are not the first of their type, but sometimes there is a decided advantage in not being first. You avoid the mistakes of others. Pioneers often get scalped.

In Cushion-Air Controls the operator will find a highly responsive system—effortless, smooth load control; Fail-Safe design—assures safety even if air pressure drops below safe operational level; Air Capacity up to twice that of any other rig—assures adequate pressure for any combination of operations; a wide range of adjustment that permits adjusting metering characteristics to suit the job.

Cushion-Air Control is a tailored system—rugged, heavy duty. Clutches and brakes are design matched to valves and cylinders—we didn't just replace a mechanical lever with an air cylinder. There are other features about this system every operator should know about. Yardage reports are increasing daily and it means greater opportunity for increased production on the part of the operator. A man at ease means more output!

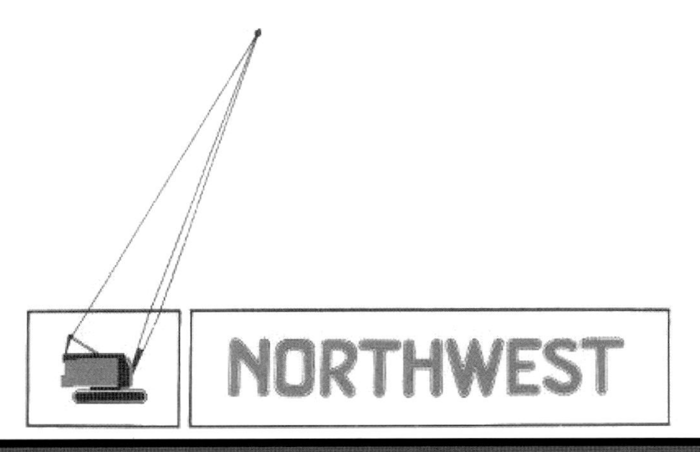

The first machine series to receive the new style of operator controls was the Model 180-D. Air assisted controls began sweeping the friction crane and excavator field in the late 1950s.

The first production Model 180-D being readied for shipment to Vecellio and Grogan Inc. of Beckley, West Virginia. The signboard by the machine's mid-boom proudly announces the shipment.

Model 180-D shovels doing what they were designed to do. "DIG ROCK!!"

For the heavy highway contractor of the mid 1960s, if you had a rock cut to do or just moving vast amounts of material to load out, a Model 180-D shovel with a good operator was hard to beat.

A Model 180-D working in a granite quarry.

A classic view of a Northwest Model 180-D shovel in a rocky load-out situation. This image would be featured many times over the years, both in Model 180-D catalog literature and ad placement in Northwest's own Material Handling Illustrated and many industry trade journals.

Three sequential photographs of a Model 180-D shovel working for Barter-Potashnick (a joint venture) on a 5.63 mile contract for I-71 through Louisville, Kentucky in 1966. The machine is seen straddle excavating a trench circa 1915 style methods in the middle photo. What would Paul Burke think?

Two views of a Model 180-D equipped with hi-front stripping shovels. With a forty foot boom and a dipper stick length of thirty-five feet, six inches and carrying a four cubic yard dipper, it was an ideal stripping machine for removing overburden.

Another Model 180-D hi-front stripper shovel.

A Model 180-D shovel equipped with a fully elevated Hi-Vue cab

Mr. Corino Civetta, owner and President of Corino Civetta Construction Corporation, New York, at 93rd Street and Second Avenue in upper Manhattan where his 180-D is excavating for a high rise building. In business for 48 years, Mr. Civetta was the first contractor to use a 180-D pullshovel in the New York area. He has been a Northwest owner since 1925 and over the years has had 28 Northwests.

Not all Model 180-D were shovels. A few went out to customers with pullshovel front-ends. The shorter and narrower undercarriage was desired by contractors like Civetta Construction Corporation of New York City who needed the big brute power and size that a 180-D could offer but had a cramped environment to work in. The pullshovel not only dug and loaded out ten-wheel dump trucks with shot rock, but set blasting mats and was used as a handy crane for pouring concrete footings, working with a lay-down type concrete bucket. The above customer testimonial appeared in the 4th quarter 1973 edition of Northwest's *Material Handling Illustrated.*

"A boy could run this Rig"

That's the statement of the operator on this 180-D in the basement excavation for one of New York's new 40-story skyscrapers. The contractor is Civetta Excavating Co. of the Bronx. Mud, rock, and the ancient wood cribbing of old docks were problems the big pullshovel had to overcome.

The 180-D is the 25th Northwest the Civetta Organization has owned. On jobs all over the country we hear of the ease of operation that comes with Cushion-Air. There is no control on any machine to equal the smoothness and response of Cushion-Air Control. It is a prime influence for the high output records Northwests make. It means the easier handling of rock, easier spotting to the truck, reduced spillage and clean-up and more truck loads per hour.

Operators who have been on Northwests with Cushion-Air never want to change. They know they can outdig anything of equal capacity.

Cushion-Air means money for you and a day of full production for the operator. There are many more Northwest advantages that assure higher production. A Northwest Sales Agent can give you the story or write the Northwest Engineering Company, 135 South LaSalle Street, Chicago, Illinois, 60603.

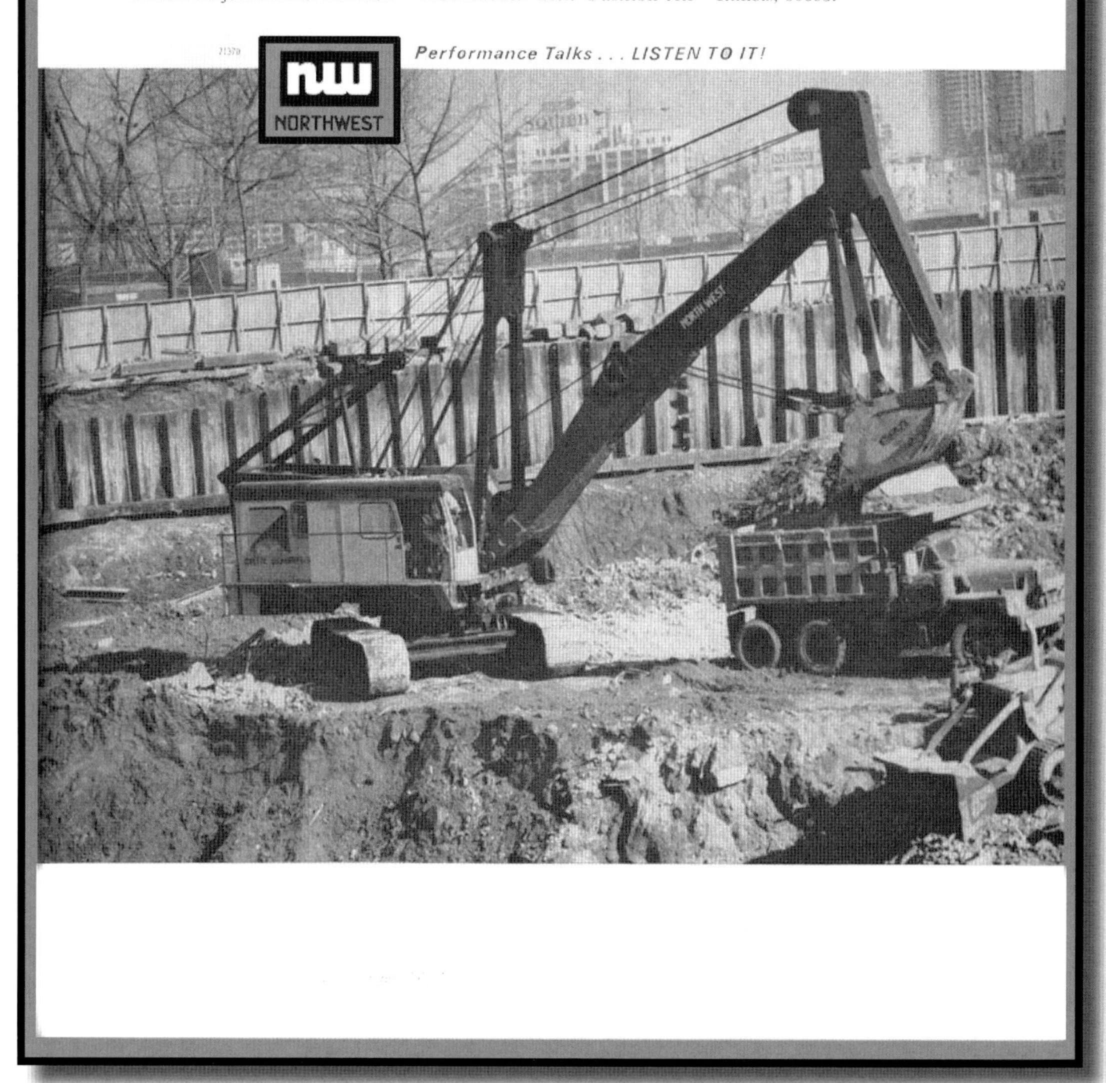

"A boy could run this rig." Civetta Excavating Company of the Bronx was also featured in many print ads of the period for Northwest. Again their new Model 180-D pullshovel is shown as an easy machine to run with its Cushion-Air controls. Northwest also pointed out that this machine was the 25th machine owned by Civetta, emphasizing repeat orders for their products. The Mack B-81 dump truck being loaded is dwarfed by the big pullshovel.

The Model 190-D

Produced from 1962 to 1969 (Series I)

The first Model 190-D was fitted with a crane boom. This photograph was taken on October 21, 1961 at the Green Bay factory. The 190-D was similar to the 180-D, but featured a longer undercarriage for stability. Its internal machinery was identical to the 180-D.

Author's correction note: Upon additional information discovered after the publication of the author's previous works regarding Northwest, it has been determined that the first Model 190-D was equipped as a crane and not a pullshovel. The above photograph was taken in the Northwest yard on the date indicated making the appearance of this machine model outfitted as a crane approximately eight months previous to the appearance of a Model 190-D outfitted as pullshovel.

The first Model 190-D pullshovel prototype. These photographs were taken on May 14, 1962. *(Two bottom photos courtesy of Dan Dobat)*

The first production Model 190-D pullshovel was delivered to Interstate Construction Inc. One of the many Ralph Cestone companies of West Orange, New Jersey, where it first worked on a heavy sewer job on the east coast of the United States. The top view shows the machine at the Green Bay factory during its predelivery run–through, dated Octobet 25, 1962. This early machine had a tubular shear-leg, tubular pitch braces installed in front of the dipper stick, and the hoist cable dead-ended on the boom mid-point.

Later production Model 190-D pullshovels featured redesigned shear leg and pitch braces of stouter construction, and relocation of the dead-ended hoist cable. This big sewer job features no less than five Northwest machines in evidence in the lower picture.

The 190-D on the Clinton Sewer contract working to 29-ft. A 41 in the background handles spoil.

This is a section of the 86-in. pipe on part of the Howard line. The pipe looks far down from the operator's seat. The cut is about 29-ft.

The 190-D on the Howard Contract starts a section of 86-in. pipe into the trench.

How it's done. Three action photographs from a *Material Handling Illustrated* of a Model 190-D pullshovel doing what it was designed to do; dig a big deep trench, and set big pipe.

Two big 190-D pullshovels working in the Detroit area. The unit on the left belonged to Rocco Ferrara, the pullshovel on the right was owned by Portland Construction. The 1960's saw a tremendous amount of large sewer and pipeline work. The Model 190-D pullshovel would prove to be a top performer when it came to moving mass quantities of deep-trench materials of any type.

This is the feature photograph to a Northwest *Material Handling Illustrated* magazine article pertaining to E & E Hauling of Bloomingdale, Illinois working on the Mallard Lake landfill reclamation project. This machine was shipped with a 110 foot boom and a 5 cubic yard bucket.

Previous page: The first Model 190-D dragline went to a customer in Altamont, California. The longer, wider crawler base made for a steady undercarriage for dragline work.

Another 110 foot boom equipped Model 190-D dragline casts material onto a spoil pile for yet another landfill project. This time the dependable 190-D was used to cut an 18-foot deep by 115 foot wide trench working 12 hour days near Joliet, Illinois.

The above scenes show the loading out a Model 190-D onto a flatbed railcar from the factory on April 25, 1963. Most 180-D and 190-D models left the Green Bay factory via rail. This would prove to be the most practical and cost-effective method to get the large machines to customers.

HENNES
FERD J. ROBERS CO.
BURLINGTON, WIS.
PHONE 763-6200

FERD J. ROBERS CO.
BURLINGTON, WIS.
PHONE 763-6200

FERD J. ROBERS CO.
BURLINGTON, WIS.
PHONE 763-6200

The final and often used method to ship a new Model 180-D or 190-D from the factory at Green Bay was to utilize the plant's own docking facilities located right on the Fox River. Shipped either by rail, truck & trailer, barge or freighter, a newly built Model 180-D or 190-D could reach any customer worldwide.

Previous page: A large marketing strategy for Northwest with their new Model 180-D and 190-D lines was the relative ease for transport via truck tractor / lowboy trailer hauling. Most times a Model 180D or 190D would arrive via rail and the machine would have to be loaded onto lowboys for delivery to their final destinations. Here, a machine for Fred J. Robers Company rides on a lowboy/jeep combination pulled by an Autocar tractor, without its crawler undercarriage and gantry folded down. A fairlead evident on this machine indicates that the customer ordered it for dragline use.

ANGELO DIPONIO
NORTHWEST

ANGELO DIPONIO
NORTHWEST

This view, and the three from the preceding page, show how Northwest utilized the equipment they manufactured for yard load-out to customers. A newly assembled Model 190-D often was found tackling the heavier lifting assignments at the Green Bay yard. The top two photographs on the previous page have the 190-D loading the upperworks of another newly manufactured machine from a wagon onto a flatcar, while the lower picture shows loaded components on a flatcar and gondola. The above picture has the 190-D loading a machinery house.

An excellent detailed photograph of a Model 190-D crane equipped with the factory optional pneumatically raised "Hi-Vue" cab. After raising the cab using compressed air, access was provided by the hinged ladder leaning against the catwalk railing. Such a machine might find application on a dockside job.

Following page: A special Model 190-D outfitted with a 190 foot boom. How appropriate that a Model 190-D be fitted with a 190 ft. boom. These series of photographs were taken in April of 1963 from the Walnut Street Bridge and the east bank of the Fox River. The 190-D with this boom looks mighty impressive. Note the mid-boom suspension required for the long boom.

WISCONSIN ACCREDITED
CHEESE COMPANY

NELSON

A very rare Northwest Model 190-D outfitted with a hi-front stripping shovel attachment. This attachment normally would go on a Model 180-D base machine but Tackett and Manning Coal Mining Corporation ordered this 190-D new from the factory in the early 1970s. The machine is seen stripping overburden at the Milestone and Longfork mine near Jenkins, Kentucky with a 4 cubic yard bucket that was the standard factory option. It is not known how many Model 190-D machines were shipped from the factory with this option or if any were bought with a long-boom or pullshovel attachments and later converted. According to Northwest special projects engineer Alan Brickett, at least one Model 190-D was sent out of the factory with the hi-front shovel attachment and a Hi-Vue cab option. The dipper had a 6 cubic yard heavy duty bucket supplied and the machine was outfitted with more counterweight similar to a dragline equipped machine. The gantry was straightened with heavier front compression struts to take on the heftier workload. This is a true testament to the original engineering and design of the model that such larger demands could have been placed on the machine without any issues. This particular machine Mr. Brickett speaks of undoubtedly would have been the largest and heaviest Northwest shovel ever produced. Only the later introduced Model 180-D Series II shovel equipped with a 10 cubic yard coal loading dipper would have had a larger capacity.

This image and the two on the proceeding pages show what would be the prototype "DA" pullshovel. This Model 190-D working for Pamco of Seattle, Washington was equipped with a one-off custom-built "DA" (dipper articulated) pullshovel. Pamco approached Northwest with a request to have such a unit built. The company had Mr. Robert Burkhart, who was the head engineer of the Model 180-D / 190-D program, design the massive articulating front-end. The stick utilized twin hydraulic rams to achieve the desired bucket curl breakout force needed. A very unique feature to the Esco supplied bucket is the side mounted rippers along with the upper bucket extensions. It is not known if these were factory supplied or field installed by the Pamco's maintenance crew.

A close up of the previous page photograph showing the details of the massive pullshovel front-end attachment with its custom articulated stick and dipper. No doubt the engineering, design and manufacture that went into this prototype "DA" pullshovel would pave the way for one of the greatest designs for below grade and trench digging ever put forth, the Northwest Model 190-DA pullshovel. Paul Burke would certainly be impressed to see how far his ideas had progressed.

Note the hydraulic excavator in the background center on this and the two previous pages is a Lorain L48H, first introduced in 1967. It was a very modern looking and operating machine for the day. Also note the man straddling the stick hydraulic cylinder untethered, another OSHA no-no in today's heavy construction field.

Appendix

The Classic year's diesel engines offerings

Two views a Fairbanks-Morse Diesel. This in-line 5 cylinder diesel, with very unusual use of individual exhaust stacks for each cylinder, is shown in these Northwest factory photographs, dated 1938. The radiator fan is mounted outboard of the radiator. Some early Model 80s carried this engine option.

The Fairbanks-Morse 5 cylinder diesel as seen from an overview view. The layout of the exhaust is clearly shown here. Also note the individual high cylinder heads. These large-bore engines ran at relatively low speed.

Northwest used a variety of diesel engine manufacturers throughout the years as either factory options or as customer requested installs. The Fairbanks-Morse diesel shown above was a rather uncommon pairing to a Northwest machine. More common pairings with other diesel engine suppliers were Atlas, Cummins, Caterpillar, General Motors, International Harvester, and the most famous name of them all, Murphy Diesel.

When one thinks of Northwest Engineering Company the name of Murphy Diesel is automatically associated with this great machine maker. The combination of these two separate entries in the mid 1930s would be a *"marriage that was truly made in heaven"*.

The Northwest Engineering Company booth at the 1936 road show in Cleveland Ohio showing a new Murphy installed in the upper rotating base of a Model 70.

A mid 1950s line card advertisement for Murphy Diesel Engines.

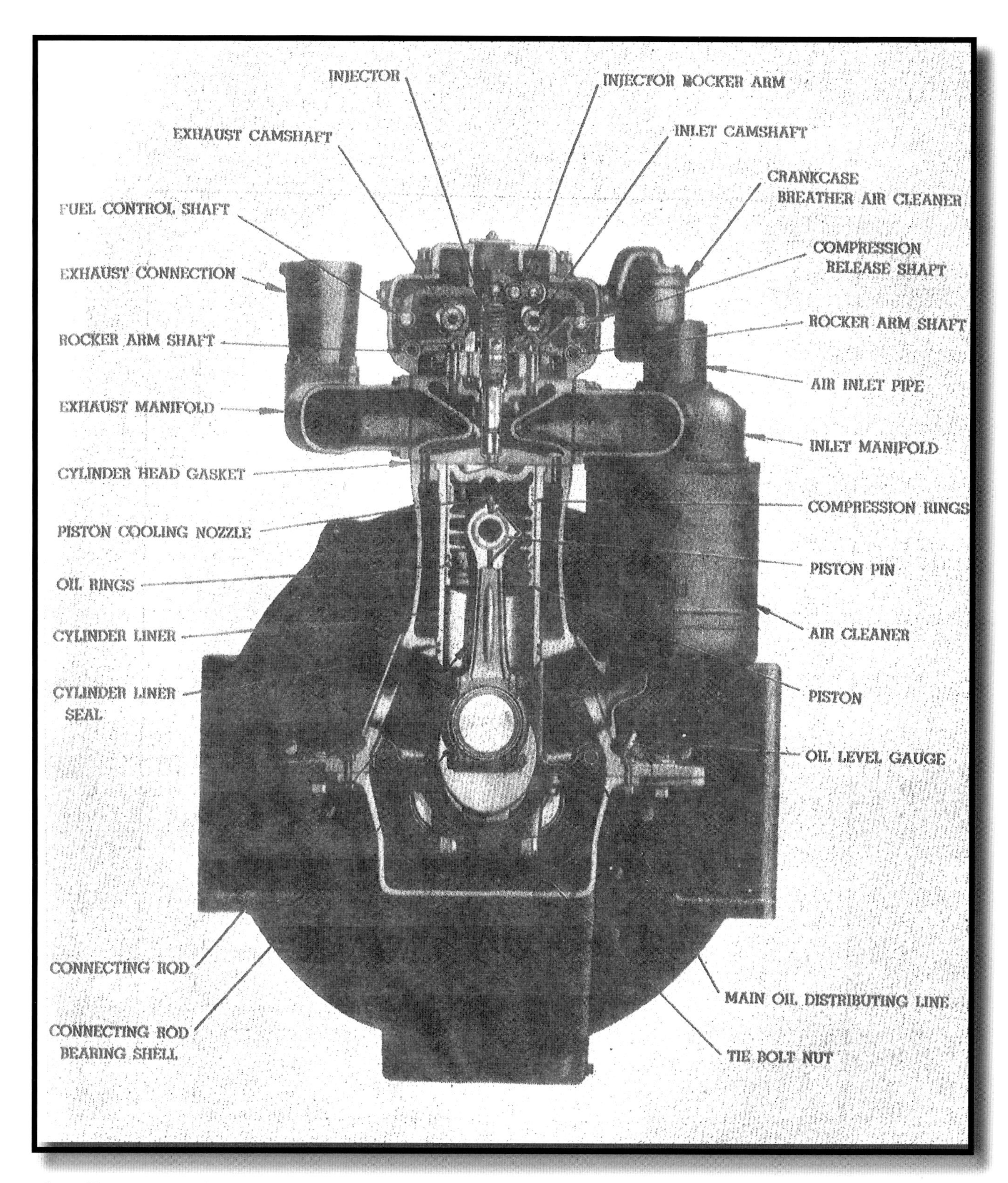

This illustration shows a cross-section drawing for the internal workings of a Murphy diesel engine. What made a Murphy a great diesel engine for a Northwest machine was its 4 valve overhead camshaft design with large rocker arms, and large cylinder displacement. All of this heavy duty engineering and design resulted in a slow lugging engine perfectly paired for a Northwest. Its reliability was well-proven in the field.

The infamous Murphy MP-21

The Murphy MP-21 in-line 6 cylinder diesel engine was one of the most commonly installed power plants in a large Northwest machine. For the Models 80-D and 95 it was the main engine of choice, but a few Model 6 machines were provided with an MP-21 instead of the more commonly supplied MP-20. The MP-21 had a bore of 5 3/4 inches and stroke of 6 1/2 inches, and delivered 170 horsepower at 1300 RPM.

177 HP DUAL FUEL RATED OUTPUT

Number of Cylinders	6
Bore and Stroke (inches)	5¾x6½
Piston Displacement (cubic inches)	1013
Governed Full Load R.P.M.	1200
Minimum Speed for Continuous Operation	800
Low Idle Speed R.P.M. (not to exceed 5 min. duration)	400

LUBRICATING SYSTEM—Full force feed oil pressure to all parts of engine. Water cooled tubular oil cooler.

OIL PUMP—Gear type. Precision cut gear teeth.

LUBRICATING OIL FILTER—Full flow in-line type with four (4) absorbent type filtering elements of large capacity. Automatic by-pass for low temperature starting.

DUAL FUEL SPECIFICATIONS

Model MP-21G

FUEL OIL FILTER—Full flow in-line high flow capacity absorbent type.

FUEL OIL TRANSFER PUMP—Precision gear type. Electrically operated for priming fuel system.

COOLING SYSTEM—Very high capacity centrifugal pump delivering water at a high velocity to all parts of cylinder block and cylinder head. Large radiator capacity with large tubes and wide fin spacing to prevent stoppage.

THERMOSTATS—Precision by-pass type thermostats which are not affected by pressure, for long engine life.

AIR CLEANER—Large, efficient oil bath type.

CRANKSHAFT—Drop forged, heat treated, alloy steel. Mirror finish induction hardened journals drilled for connecting rod lubrication and piston cooling.

CRANKSHAFT MAIN BEARINGS—Seven precision steel-backed bearings lead-tin plated for protection against corrosion.

Main Bearing Diameters and Lengths in Inches	
Front	4 x 2⅞
Center	4 x 4¼
Intermediate	4 x 2¼
Rear	4 x 3 15/16
Connecting Rod	4 x 2¾

CONNECTING RODS—Forged heat treated alloy steel drilled for pressure lubrication to steel backed, diamond bored wrist pin bushings and for piston cooling.

CRANKCASE—Cylinder block integral with crankcase, symmetrically designed to maintain perfect alignment. Heat treated alloy steel tie studs securing main bearing caps and cylinder head.

CRANKCASE VENTILATOR—Fresh air circulation type. Air taken in at camhousing through the breather air cleaner and taken out at crankcase.

CYLINDER LINERS—Replaceable wet type made from alloy castings of high hardness with honed oil retaining finish.

CAMSHAFTS—Two overhead. Drop forged, with integral cams. Hardened and ground. Seven steel back, copper-lead precision bearings. Rifle drilled for pressure lubrication to all bearings.

CAMSHAFT DRIVE—Through bevel gears and worm gears. Drive is taken off of crankshaft at flywheel end to secure uniform rotation. Full pressure lubrication.

VALVES—Four alloy steel overhead valves per cylinder located symmetrically around each injector. Special alloy replaceable valve seat. High grade alloy valve guides. (Four valves for optimum valve area for good intake and exhaust.)

VALVE OPERATION—Direct from camshafts through heat treated rocker arms with hardened faces.

GOVERNOR—Murphy Diesel full variable speed hydraulic servo-type governor. At reduced engine speed, when required, the governor will open the throttle in response to load demand.

SAFETY CONTROL—Engine automatically stops if lube oil pressure falls below a safe minimum.

PISTONS—Oil cooled type made from fine grained, heat treated castings, precision machined and tin plated. Pistons fitted with high quality piston rings.

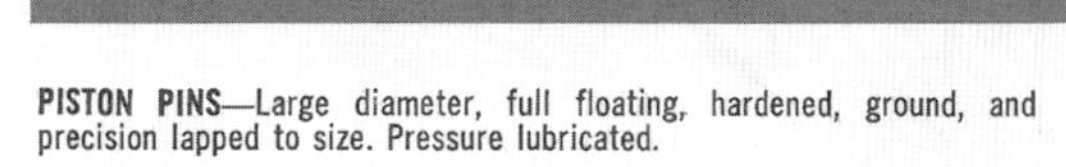

PISTON PINS—Large diameter, full floating, hardened, ground, and precision lapped to size. Pressure lubricated.

FLYWHEEL AND FLYWHEEL HOUSING—S.A.E. No. 0, flywheel made from high tensile material.

ELECTRIC STARTING SYSTEM—24 volt high torque starting motor, geared directly to ring gear on flywheel. Manually operated gear engagement and switch operation. Ball bearing generator and voltage regulator.

UNIT INJECTORS—Murphy Diesel design, combining fuel pump, precise metering control—and spray nozzle into a single unit. One per cylinder. On dual fuel operation serves to inject pilot fuel at constant throttle setting (see curves).

INJECTOR DRIVE—Through anti friction roller bearing from overhead camshaft.

COMBUSTION CHAMBER—Shape of combustion chamber gives lower fuel consumption, because more heat is utilized in power.

DUAL-FUEL SYSTEM—Designed to use stripped natural gas ignited by pilot fuel oil injection. A simple lever shifts operation to fuel oil when desired. If permanent operation as a full diesel engine is desired for continuous diesel service, this change can be made in a short time without any additional parts.

FUEL SPECIFICATIONS—Natural gas (Methane) stripped of Butane, Propane, etc., essentially free of air and other foreign material; less than 10 grains of sulphur per 100 cu. ft. of gas; heat content approximately 900 to 1100 B.T.U. per cu. ft.

GAS VALVES AND REGULATORS—All necessary valves and regulators are furnished: metering gas valve, zero pressure governor, reduced pressure regulator, gas strainer and shut off cock.

INSTALLATION DIMENSIONS—Shown on drawings.

Murphy Diesel Company reserves the right to discontinue models or change specifications, material, design, or prices at any time without notice and without incurring any obligation.

A Murphy spec sheet for an MP-21G, a dual-fuel engine. The dual-fuel MP-21G was designed to run on either natural gas or diesel fuel. The engine shown here is similar to the MP-21. The MP-21 used in a Northwest 80-D or Model 95 operated on diesel fuel only. The engine shown is mounted on a skid.

SPECIFICATIONS MODEL 124

Number of Cylinders	6
Bore and Stroke (inches)	6⅜x6½
Piston Displacement (cubic inches)	1245
Governed Full Load R.P.M.	1200
Minimum Speed for Continuous Operation	800
Low idle speed R.P.M. (not to exceed 5 min. duration)	400
Continuous brake H.P. without Fan at 1200 R.P.M.	232
Continuous brake H.P. with Fan at 1200 R.P.M.	225

Main Bearing Diameters and Lengths in Inches

Front	4x2⅞
Center	4x4¼
Intermediate	4x2¼
Rear	4x3⅛
Connecting Rod	4x2¾

MURPHY DIESEL

LUBRICATING SYSTEM—Full force feed oil pressure to all parts of engine. Water cooled tubular oil cooler.

OIL PUMP—Gear type. Precision cut, and hardened gear teeth.

LUBRICATING OIL FILTER—Full flow in-line type with four (4) absorbent type filtering elements of large capacity. Automatic by-pass for low temperature starting.

FUEL FILTER—In-line high flow capacity absorbent type filter.

FUEL TRANSFER PUMP—Precision gear type, driven from generator shaft. Electrically operated for priming fuel system.

COOLING SYSTEM—Very high capacity centrifugal pump delivering water at a high velocity to all parts of cylinder block and cylinder head. Large radiator capacity with large tubes and wide fin spacing to prevent stoppage.

THERMOSTATS—Precision by-pass type thermostats for long engine life.

AIR CLEANER—Large, efficient oil bath type.

CRANKSHAFT—Drop forged, heat treated, high quality forging steel. Mirror finish, Tocco hardened journals drilled for connecting rod lubrication and piston cooling.

CRANKSHAFT MAIN BEARINGS—Seven precision trimetal type bearings, steel back, copper, lead, nickel, and lead-tin plated to minimize corrosion.

CONNECTING RODS—Heat treated, drop forged, high quality forging steel drilled for pressure lubrication to steel back bronze lined wrist pin bushings and for piston cooling.

CRANKCASE—Cylinder block integral with crankcase, symmetrically designed to maintain perfect alignment. Heat treated alloy steel tie bolts securing main bearing caps and cylinder head.

CRANKCASE VENTILATOR—Fresh air circulation type. Air taken in at camhousing through the breather air cleaner and taken out at crankcase.

CYLINDER LINERS—Replaceable wet type made from alloy castings of high hardness with mirror finish.

CAMSHAFTS—Two overhead. Drop forged, with integral cams. Hardened and ground. Seven steel back, copper-lead precision bearings. Rifle drilled for pressure lubrication to all bearings, journals and cams.

CAMSHAFT DRIVE—Through bevel gears and worm gears. Drive is taken off of crankshaft at flywheel end to secure uniform rotation. Full pressure lubrication.

VALVES—Four alloy steel overhead valves per cylinder located symmetrically around each injector. Special alloy replaceable valve seat. High grade bronze valve guides. (Four valves for optimum valve area for good intake and exhaust.)

VALVE OPERATION—Direct from camshafts through hard faced, forged rocker arms.

GOVERNOR—Murphy Diesel full variable speed hydraulic servo-type governor. At reduced engine speed, when required, the governor will open the throttle in response to load demand.

SAFETY CONTROL—Engine automatically stops if lube oil pressure falls below a safe minimum.

PISTONS—Oil cooled type made from fine grained, heat treated castings, precision machined and tin plated. Pistons fitted with high quality piston rings.

PISTON PINS—Large diameter, full floating, hardened, ground, and precision lapped to size. Pressure lubricated.

ELECTRIC STARTING SYSTEM—24 Volt high torque starting motor, geared directly to ring gear on flywheel. Manually operated gear engagement and switch operation. Ball bearing generator and voltage regulator.

UNIT INJECTORS—Murphy Diesel design, combining fuel pump, precise metering control and spray nozzle into a single unit. One per cylinder.

INJECTOR DRIVE—Through anti-friction roller bearing from overhead camshaft.

COMBUSTION CHAMBER—Shape of combustion chamber gives lower fuel consumption, because more heat is utilized in power.

TUBO CHARGER—Murphy special designed power input and combustion system. Greatly increases power without added wearing parts.

INSTALLATION DIMENSIONS—Shown on drawings.

A spec sheet for the Murphy Model 124 which was a 6 cylinder diesel used for a brief period of time in the Model 180-D prior to the introduction of the more common Model 872 V-8.

A 4-cylinder Murphy MP-11, these were used extensively in the Northwest Model 25-D throughout the years. The engine was rated at 65 flywheel horsepower @ 950 RPM with a 5-3/4 bore and 6″ stroke.

Period Northwest Engineering Company

Road Show photographs

The Northwest booth at the 1936 Road show in Cleveland.

The 1957 Road show. Show-goers inspect the offerings by Northwest.

Views of the 1963 Road Show at the International Amphitheatre in Chicago, Illinois.

An overview of the exhibits at the show from the Northwest corner.

Prospective customers crowd around the new 180-D shovel at the 1963 exhibition. A viewing platform was built around the machine for easy inspection by prospective new owners.

The 1969 Road show: the future

This photograph taken at the 1969 Road Show is a few years past what the authors consider the classic years for Northwest. The year 1969 probably was the peak year for heavy construction in the United States, with the Interstate Highway program in full swing. This image depicts the changes in machine design that was occurring at end of that era, especially evident by the number of hydraulic excavators displayed on the exhibition floor.

In the foreground center, the newly introduced Model 70-D crane with independent hydraulic swing can be seen, along with the new Model 30-DH convertible hydraulic excavator to the left. These new offerings from Northwest were a major departure in both motive power and function for a machine. The company at this time was clearly embracing new technology, not by choice, but by necessity.

Louis E. Houston and his sons

With the death of Louis E. (L.E.) Houston (sitting) on September 6, 1966, the helm of Northwest Engineering Company would transfer to his eldest son Allen F. Houston (standing left). L.E's tenure at Northwest as the head of the company starting in 1926 saw a tremendous change in the construction equipment manufacturing industry. During his reign, small independent enterprises became larger, refined, and specialized enterprises with a corporate nature that would have engineering talent of the likes never before seen.

His other two sons, Kenneth (standing center) and Robert (standing right) would play roles in his business empire. Robert would be set up to run Murphy Diesel in Milwaukee. Kenneth would for a time be located to the west coast of United States to play a role in the Northwest's sales organization. He would return to Green Bay in the mid- 1950s to head up the company's personnel department.

The end for now.

For a more comprehesive history on Northwest Engineering Company, its products, and key personel; the following title is available by the author's

ISBN: 9781494342012

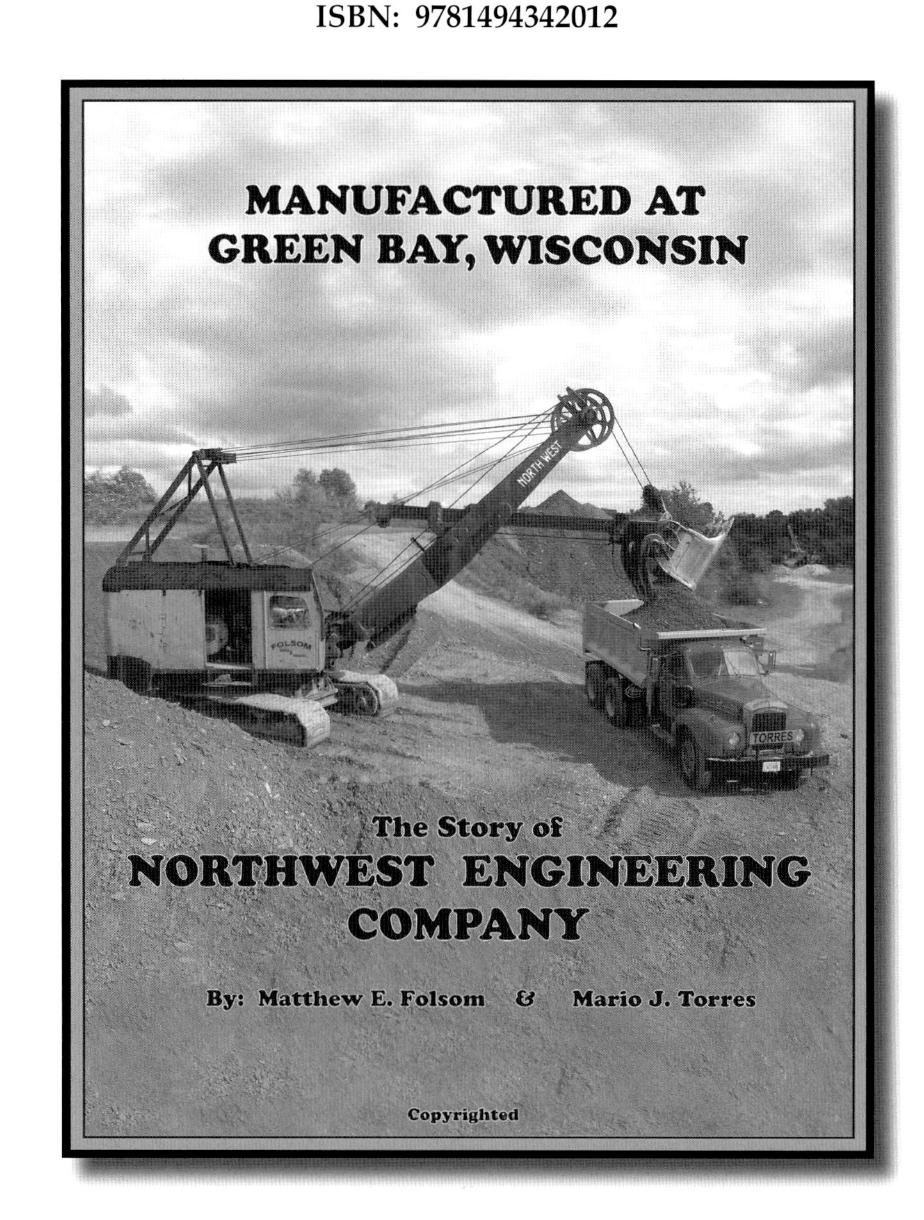

"Manufactured at Green Bay, Wisconsin"
is the account of Northwest Engineering Company. From its early days as a builder of tugboats for the World War One war effort to the role the company played as one of the premier manufacturers of excavators in the world. The team of Folsom and Torres trace Northwest's rise and eventual demise with vivid clarity giving an account of the company's key personnel and products. With an informative text and over 200 Black & white and color images from the author's and past employee's personal collections, heavy equipment enthusiasts and devotees to the history of local interests and the heavy construction machinery industry will find new insight into this past enterprise of Titletown, U.S.A.